Geoinformatics Applications in Agriculture

AF638161

NIPA® GENX ELECTRONIC RESOURCES & SOLUTIONS P. LTD.
New Delhi-110 034

Geoinformatics Applications in Agriculture

Editors :

ANIL KUMAR SINGH

Project Director, Water Technology Centre

&

USHA KIRAN CHOPRA

Principal Scientist, Division of Agricultural Physics

Indian Agricultural Research Institute, New Delhi

NIPA® GENX ELECTRONIC RESOURCES & SOLUTIONS P. LTD.

New Delhi-110 034

NIPA® GENX ELECTRONIC
RESOURCES & SOLUTIONS P. LTD.

101,103, Vikas Surya Plaza, CU Block
L.S.C. Market, Pitam Pura, New Delhi-110 034
Ph : +91 11 27341616, 27341717, 27341718
E-mail: newindiapublishingagency@gmail.com
Web: www.nipabooks.com

For customer assistance, please contact
Phone: + 91-11-27 34 17 17
Fax: + 91-11- 27 34 16 16
E-Mail: feedbacks@nipabooks.com

ISBN: 978-81-19215-45-4

Composed and Designed by NIPA®.

M S SWAMINATHAN RESEARCH FOUNDATION

M S Swaminathan
Chairman

3rd Cross Street, Taramani Institutional Area
Chennai (Madras) - 600 113 (INDIA)
Phone : +91-44-22641698, 22542790
Fax : +91-44-22541319
E-mail : chairman@mssrf.res.in. swami@mssrf.res.in

FOREWORD

The most urgent test facing our agriculture is the need for the enhancement of the productivity and profitability of major farming systems on an environmentally sustainable basis. Several years ago I coined the term "Ever-green revolution" to indicate the need for improving productivity in perpetuity without associated ecological harm. The Ever-green revolution technology requires the availability of spatially referenced extensive data base on soil, water and cover, forests and socio-economic information. This is where the field of Geoinformatics is extremely helpful. Geoinformatics is the scientific congiomeration of remote sensing, geographic information system, cartography, photogrammetry and global positioning system and has virtually limitless potential through a range of applications as described in this book.

It is an act of vision on the part of Government of India to have set up a Ministry of Earth Sciences. This book therefore is a timely contribution. It will be a very useful reference book to scientists, scholars and policy makers. I congratulate Dr. A K Singh and Dr. Usha Kiran Chopra for their writing efforts to bring out this very valuable publication.

November, 2006

M S Swaminathan

PREFACE

The good to keep up with the demands of the ever increasing population is necessitating development of technologies which can retrieve, process and disseminate information in an effective and reliable manner at the fastest pace possible. The rapid development of information and communication technologies (ICT) is, therefore, going to play a very vital role in the development and progress of any organization or nation. Realizing the importance of this, a Symposium on *"Geoinformatics Applications for Sustainable Development"* was organized at IARI, New Delhi to have in-depth deliberations on the state-of-art of various tools that can be used for evaluating, monitoring, managing and planning all aspects which are essential for all development processes. The writing of this book was conceived during that period. Since geoinformatics has now become very important as a tool, it was felt necessary to bring out a publication examining the scope of this important science.

Earth functions as a complex system and existing infrastructure and methodologies are inadequate for the community to address many problems of this system. It requires an integrated and innovative approach to analyse, model and develop extensive and diverse data sets for the solutions of these problems. Currently, there is a chaotic distribution of available data sets, lack of documentation about them and the methodologies, lack of easy-to-access tools, computer modeling etc. which pose the major obstacles for the earth scientists in trying to work out solutions. Geoinformatics is a rapidly developing science which encompasses all modern tools and techniques like remote sensing, GIS, GPS, photogrammetry along with computer modeling to systematic data storage, analysis and retrieval for research as well as for documentation to bring out relevant information and advancements in various fields.

This book presents a comprehensive account of various topics related to the applications of geoinformatics which are highly diversified and include space and environmental monitoring, watershed management, water resources and management, productivity enhancement in agro-ecosystems, resource information systems, precision farming, monitoring of glaciers and lakes, geospatial techniques and livelihood improvement using GIS.

We hope that this book will provide the readers with a basic understanding of what Geoinformatics is all about and the future scope of this modern tool.

November, 2006 **Editors**

List of Contributors

A. SARANGI
Scientist (SS)
Water Technology Centre
(I.A.R.I.), New Delhi - 110 012

A.K.BHATTACHARYA
Water Technology Centre
(I.A.R.I.), New Delhi - 110 012

A.K. SIKKA
ICAR Research Complex for
Eastern Region, WALMI Complex
Phulwari Sharif, Patna
Bihar - 801 505

A.K. SINGH
Project Director
Water Technology Centre
(I.A.R.I.), New Delhi - 110 012

AJAI
Space Applications Centre
(I.S.R.O.), Ahmedabad - 380 015

ANSHU DIXIT
I.A.S.R.I., Library Avenue
New Delhi - 110 012

B. SAHA
ICAR Research Complex for
Eastern Region, WALMI Complex
Phulwari Sharif, Patna
Bihar - 801 505

BASANTA SHRESTHA
International Centre for Integrated
Mountain Development
Kathmandu, Nepal

C.B.S. DUTT
Indian Space Research
Organisation (I.S.R.O.)
Headquarter, Bangalore - 560 094

CHITRA MOTHKUR
Indian Institute of Horticultural
Research (IIHR), Hessaraghatta
Bangalore

CHITRA SOOD
Centre for Geo-informatics
Research and Training, CSK
Himachal Pradesh Agriculture
University, Palampur - 176 062

D. CHAKRABORTY
Division of Agricultural Physics
(I.A.R.I.), New Delhi -110 012

D.K.SINGH
Water Technology Centre
(I.A.R.I.), New Delhi-110 012

D. SREENIVASA MURTHY
Indian Institute of Horticultural
Research (IIHR), Hessaraghatta
Bangalore

G.S. GEETHA
Indian Institute of Horticultural Research (IIHR), Hessaraghatta
Bangalore

H.C. MEENA RANI
Indian Space Research Organisation (I.S.R.O.)
Headquarter, Bangalore - 560 094

I.P. ABROL
Centre for Advancement of Sustainable Agriculture
NASC Complex, New Delhi

J.K. GARG
Space Applications Centre (I.S.R.O.), Ahmedabad - 380 015

K. D. SHARMA
National Institute of Hydrology
Roorkee- 247 667

L.B. ASHOK
Indian Institute of Horticultural Research (IIHR), Hessaraghatta
Bangalore

M.S. MANE
Dr.B.S.Konkan Krishi Vidhyapeeth
Dapoli, Ratnagiri

N. KALRA
Division of Agricultural Physics
(I.A.R.I.), New Delhi -110 012

P.K. MALHOTRA
I.A.S.R.I., Library Avenue
New Delhi - 110 012

P.R. BHATNAGAR
ICAR Research Complex for Eastern Region, WALMI Complex
Phulwari Sharif, Patna
Bihar - 801 505

P.S. ROY
National Remote Sensing Agency
(Dept. of Space, Govt. of India)
Hyderabad – 500 037

PRADEEP MOOL
International Centre for Integrated Mountain Development
Kathmandu, Nepal

R.K. GUPTA
Rice Wheat Consortium
CIMMYT, NASC Complex
New Delhi – 110 012

R.K. TOMAR
Division of Agricultural Physics
(I.A.R.I.), New Delhi -110 012

R.M. BHAGAT
Centre for Geo-informatics Research and Training, CSK
Himachal Pradesh Agriculture University, Palampur - 176 062

R. N. SAHOO
Division of Agricultural Physics
(I.A.R.I.), New Delhi -110 012

RAJENDER PRASAD
I.A.S.R.I., Library Avenue
New Delhi - 110 012

R. NAGARAJA
Indian Space Research Organisation (I.S.R.O.)
Headquarter
Bangalore - 560 094

R.P.JONES
University of East Anglia
Norwitch, U.K

RAKESH KUMAR
National Institute of Hydrology
Roorkee- 247 667

REENA ROSY THOMAS
Indian Institute of Horticultural
Research (IIHR), Hessaraghatta
Bangalore

S.K. SRIVASTAVA
Indian Space Research
Organisation (I.S.R.O.)
Headquarter, Bangalore - 560 094

S. BAJRACHARYA
International Centre for Integrated
Mountain Development
Kathmandu, Nepal

S. BANDYOPADHAYAY
Indian Space Research
Organisation (I.S.R.O.)
Headquarter, Bangalore - 560 094

S. CHOWDHURY
CPSL (NGO), Anishabad, Patna
Bihar

S.C. KOTUR
Indian Institute of Horticultural
Research (IIHR), Hessaraghatta
Bangalore

SANJAY K. JAIN
National Institute of Hydrology
Roorkee- 247 667

T.M. GAJANANA
Indian Institute of Horticultural
Research (IIHR), Hessaraghatta
Bangalore

V.K.GUPTA
I.A.S.R.I., Library Avenue
New Delhi - 110 012

V.K. SEHGAL
Division of Agricultural Physics
I.A.R.I., New Delhi -110 012

V.S. HEGDE
Indian Space Research
Organisation (I.S.R.O.)
Headquarter, Bangalore - 560 094

V. JAYARAMAN
Indian Space Research
Organisation (I.S.R.O.)
Headquarter, Bangalore - 560 094

VAIBHAV KALIA
Centre for Geo-informatics
Research and Training, CSK
Himachal Pradesh Agriculture
University, Palampur - 176 062

U. K. Chopra
Principal Scientist
Division of Agricultural Physics
I.A.R.I., New Delhi - 110 012

❑❑❑

CONTENTS

GEOINFORMATICS APPLICATIONS IN AGRICULTURE, 2007
Eds. : A.K. Singh and U.K. Chopra; pp. 1-38
New India Publishing Agency, New Delhi (India)
E-mail : newindiapublishingagency@gmail.com
Web: www.bookfactoryindia.com

1

Space Sensing and Environment Monitoring

Ajai & J.K. Garg

Key words : Remote Sensing, Ecosystem, Environmental Pollution, Biodiversity Conservation, Coastal Environmental Degradation, Wetland.

Introduction

National prosperity and human resources development are primarily shaped by the availability of natural resources and their use. None the less, the benefits are quite often negated by the resultant environmental degradation concomitant with their utilisation for economic values. The problems of environmental degradation have assumed critical dimensions in many parts of India due to resources exploitation; industrialisation and urbanisation, unchecked population growth and a never ending quest for energy to meet the increasing demands. Moreover, improper utilisation of resources has resulted in such environmental problems as soil erosion, waterlogging, floods, droughts and pollution of various types besides socio-economic disparity among people.

After the 1972 Stockholm conference on Human Environment and Development, there has been a phenomenal increase in the worldwide interest in the environmental issues associated with development. Consequently, the focus of development has shifted to environment friendly and sustainable developmental planning in recent years. The Rio

Summit (June 3-14, 1992) further emphasised the fact that earth is an integral and interdependent system, which require sustainable developmental planning. During deliberations of the Rio Summit (Earth Summit), it was emphasised that satellite imagery will be quite helpful in monitoring earth's environment.

India has a long tradition of environmental protection and has enacted laws during the last three decades to give further impetus to the conservation of natural resources and protection of environment. Some of the important laws are:

- Water (Prevention and Control of Pollution) Act, 1974
- Water (Prevention and Control of Pollution) Cess Act, 1977
- Air (Prevention and Control of Pollution) Act, 1981
- The Environment (Protection) Act, 1986
- Forest (Conservation) Act, 1980
- National Forest Policy, 1988
- The Wildlife (Protection) Act, 1972

Major Environmental Issues in India

Environmental problems in India can be categorised in three categories. Type I problems are due to lack of development such as deforestation for fuel wood, timber extraction, filling of wetlands, soil erosion, unhygienic slums, absence of sanitation and safe water supply etc. Type II environmental problems arise due to lack of environmental safeguards in the developmental projects such as air and water pollution, hazardous waste disposal, pesticide pollution, fertiliser run-off into streams, waterlogging, etc. Type III problems are of global nature (global warming, ozone depletion, marine pollution, loss of biodiversity etc.). India has a considerable backlog of environmental degradation. Some of the major environmental issues in the country are:

- Ecosystem degradation (wetlands, forests, grasslands, coral reefs, mangroves, islands)
- Environmental Impact Assessment (of developmental activities and urbanisation)

- Environmental Pollution (water, land, air)
- Environmental hazards (landslides, mine fires, forest fires, floods, industry etc.)
- Desertification
- Loss of Biodiversity
- Protected area conservation (Biosphere reserves, national parks, sanctuaries)
- Coastal environmental issues (mangrove degradation, coral reefs, CRZ etc.)
- Global environmental issues

Ecosystem Degradation

An ecosystem, the word coined by Sir Arthur Tansley in 1935, is defined as, "a spatially explicit unit of the Earth that includes all of the organisms, along with all components of the abiotic environment within its boundaries". In common sense an ecosystem comprises the biological community that occurs in some locale, and the physical and chemical factors that make up its non-living or abiotic environment. There are many examples of ecosystems - a wetland, a forest, an estuary, and grassland and so on so forth. In the broadest sense ecosystems are commonly categorized into terrestrial and aquatic ecosystems. They can be further classified depending on the substrate, dominant species and abiotic conditions:

Terrestrial Ecosystems		**Aquatic Ecosystems**	
Forests	Temperate, tropical, Evergreen	Fresh Water	Lentic, Lotic, wetlands, riparian, riverine, palustrine
Grasslands	Savannah, steppe, shola	Marine	Coral reefs Mangrove
Agro-ecosystems	Rainfed, irrigated	Estuarine	Bay, lagoon, creek, back water
Arid Mountain	Desert, semi-arid Himalayan, taiga, W. Ghats	Tidal	Salt Marsh

Forests

Forests are an important global resource on which human populations depend for wood, food, fodder, maintenance of air quality, recreation and many other uses. They also serve as habitats for millions of plant and animal species. Conversion of forests into uses such as agriculture, urban expansion and mining etc. and degradation of forests are the major ecological problem confronting the nation. National Forest Policy of India stipulates that 1/3rd of the country's land is required to be maintained under the forest cover. Recorded forest cover, however, is only 765210 sq. km (23.28 % of the geographical area). This includes 416516 sq. km of reserved forests, 223309 sq. km of protected forests and 125385 sq. km unclassed forests. According to Forest Survey of India during 1951-1980 about 4.328 million ha area has been lost to various non-forest uses. The major part of forest has been diverted to agriculture (2.623 Mha), river valley projects (0.502 Mha), industries and townships (0.134 Mha), transmission lines and roads (0.061 Mha). Even after the enforcement of Forest Conservation Act in 1980 an additional 0.3236 million ha of forestland has been diverted to non-forest uses. This calls for taking measures to arrest this trend and bring additional areas under forests by way of afforestation and reforestation. Other problems of forestry include biodiversity conservation, forest fires etc.

Wetlands

Wetlands are multifunctional and most productive ecosystems providing an array of benefits to the mankind. However, due to the poor perception of their functions and values they are under the constant threat of obliteration and degradation. They are not only the sources of drinking water to burgeoning human population, but also provide a myriad of benefits like flood control, ground water recharge, shoreline protection, and aquaculture habitats. In addition, they are the habitats for a variety of waterfowl. According to the first scientific inventory of wetlands of the country carried out by Space Applications Centre, Ahmedabad using IRS satellite data (Garg *et al*, 1998), wetlands occupy about 8 Mha area. The estimates do not include paddies and river/streams. As per an estimate by the Asian Wetland Bureau, area under wetlands is 58.2 Mha including paddy cultivation, which is 40.9 Mha (excluding rivers). Ministry of Environment and Forests has estimated area under wetlands to be 4.1 Mha excluding paddy fields and mangroves. Major environmental threats to the wetlands are encroachment, pollution, dredging, aquaculture, siltation and weed infestation etc.

Coastal wetlands provide a vital link in the flow of matter and energy among various components of coastal ecosystems. Coral reefs, mangroves, estuaries, tidal flats and seabed grass are the important critical habitats needing conservation and preservation. Mangroves are considered ecologically sensitive areas under Environmental Protection Act, 1986. However, due to exploitation for various human uses many areas have been obliterated or are under constant threat of obliteration. Similarly, coral reefs are hot spots of marine biodiversity. Coral reefs are shallow water tropical ecosystems characterised by remarkably high biomass production and rich floral and faunal diversity. Regular monitoring of coral reefs is required so as to protect these vital habitats.

Biodiversity

India is one of the 12-mega biodiversity centres of the world. About 46,000 species of wild plants and over 81 000 sp. of wild animals are present in India (6% of world's known wildlife). Out of these about 1500 plant and animal species are in the endangered category. Due to anthropogenic activities, which cause degradation and loss of their habitats, there is a decline in the number of species with each passing day. Concerted efforts are required to preserve our gene pool for posterity. This calls for taking remedial action in the fragile areas to further arrest the loss of species.

Environmental Impact of Developmental Activities

The Ministry of Environment and Forests took up EIA Programme in 1977-78. MoEF has made mandatory the submission of EIA before any major developmental activity can be taken up. A total of 30 such projects have been identified by the Ministry which need environmental clearance from the central govt. Developmental activities like mining, pit-head thermal power stations, hydro-power, major irrigation projects, ports and harbours (excluding minor ports) and prospecting and exploration of major minerals in areas above 500 ha, the project authorities are required to intimate the location of the project site to the Central Government in the Ministry of Environment and Forests while initiating any investigation. In the following sections environmental issues related to mining, and thermal power generation projects are discussed:

Mining

The environmental aspects of mining depend to a great extent on the type of mining, the ore to be extracted, benefaction, and the machinery used for obtaining desired concentration levels etc. The damage to the environment and ecology are difficult to assess in quantitative terms though all mining activities have their attendant problems of deforestation, air and water pollution, disfiguring of the landscape, increase in soil erosion, removal of fertile topsoil and subsidence of land etc. The working group on Mining and Environment has identified the following environmental problems of serious nature normally associated with mining activities:

- Water pollution due to wash off of overburden wastes, discharge of mine water, discharge of tailing into the water bodies, and mine drainage and coal-washing operations.
- Air pollution due to release of noxious gases into the atmosphere and generation of dust during various operations.
- Land degradation due to large-scale excavation, disposal of mine wastes and tailing and subsidence.
- Landslides in the hilly terrain due to steepening of slopes during opencast mining operations. Deforestation due to felling of trees for construction of roads, fuel wood, etc.

- Disruption of water regime.
- Problem of noise and vibration due to blasting and operation of heavy machinery.
- Human-environment problems: settlement, health and allied problems.Damage to sites of cultural, historical and scenic importance.
- Problems caused by mine fires, like land degradation, air and thermal pollution, disruption of transportation etc.

The EIA of mining activity should include baseline information on land cover, vegetation pattern, geology/ geomorphology, hydrology, soil, drainage pattern, physiography and well-defined conservation plans. EIA should also focus on likely environmental impacts on land, air and water during operational and post-operational phases.

Thermal Power Generation

Electric power plays a crucial role in the economic development. The total installed capacity of power generation, which was 1400 in 1947 MW, has gone upto 98000 MW by 1999. Out of this thermal power accounts 69 per cent, hydro 29 per cent and nuclear power about 2 per cent. Closely linked with thermal power generation are the problems of air, water and land pollution. In the coal burning process, besides gaseous pollutants (SO_2, NO_2, CO_2, HC etc.), particulate pollutants (fly ash) are also generated. In the absence of any control equipment like electrostatic precipitators, the fly ash gets deposited on the land and causes serious land pollution problems. Even in the newer power plants where it is deposited in the ash ponds, it gets transported due to wind action and causes land pollution in the surrounding areas. Problem of land pollution by thermal power generation further gets accentuated if power plants are located at coal pitheads. Major environmental problems associated with thermal power generation are:

- Air pollution due to release of obnoxious gases (SO_2, NO_x CO_2 etc.).

- Water pollution due to emission of heat and other effluents from the power plants.
- Land cover/use changes due to construction, associated industrialisation and urbanisation.
- Damage to the vegetation and other ecosystem components.
- Noise pollution due to vibration and operation of heavy machinery.
- Human environment problems (settlements, health, etc.).
- Damage to sites of cultural, historical and scenic importance due to pollution.

Environmental Pollution

The problems of environmental pollution have assumed critical dimensions in many parts of India due to industrialisation, urbanisation and a never-ending quest for energy to meet the increasing demands. Improper siting of industrial units and their localisation has resulted in pollution of various types besides socio-economic disparity among people. Delhi, Mumbai, Ahmedabad, Calcutta, Hyderabad, Vadodara, Korba, Nagpur, Chennai, Kanpur are some of the major cities in the country where air pollution problem has reached menacing proportions.

Air Pollution: Thermal power plants, refineries, smelters, automobiles and industrial installations emit enormous quantities of pollutants especially SO_2, NO_X etc. in the atmosphere resulting in acute pollution problem in many parts of the country. A super thermal power plant using even normal or low sulphur coal will emit about 100 tonnes of SO_2 everyday. Air pollution in the metro cities and selected industrial pockets of the country has reached alarming levels. Depending on the meteorological conditions these pollutants are transported to long distances and cause damage to vegetation, man, animals and various other components of ecosystems. Some of these pollutants also cause global warming due to greenhouse effect (methane, carbon dioxide, chlorofluorocarbons) and acid rain (sulphur dioxide, CO_2 etc.).

Land pollution: Out of 329 million ha of the total area of the country, 175 ha of land require special treatment to restore such lands to productive use. Out of this, water and wind erosion affected land is 150 million ha, salinity and alkalinity 8 million ha, and river action and other factors (ravines, gully erosion etc.) 7 million ha. Land area likely to be lost annually by mining is 1500 ha. Pollution caused by solid wastes is another cause of concern. India generates about 25 million tonnes of solid wastes every year. Fly ash and mining activities also result in land pollution.

Water Pollution : Health of our aquatic environment has been causing serious concern among environmentalists and public in recent years. Pollution of water has assumed serious proportions in many lentic and lotic ecosystems. In fact, 70 per cent of fresh water in the country is polluted. Water is polluted mainly by domestic sewage, industrial effluents/wastes, increasing use of fertilisers and pesticides for agricultural use and silt generated by degraded catchments. In addition, air pollution also causes water pollution many a time, as for example, lowering of pH of lakes due to acid rain etc.

Environmental Hazards

Landslides, earthquakes, cyclones; floods, forest fire, mine fire etc are the major environmental hazards that cause extensive damage every year. However, no account is available either of the affected area(s) or the damage.

Landslides are among the most devastating and frequently occurring disasters in the hilly areas, especially the Himalayas, causing colossal loss of life and property. Though, the exact cause of landslide is not fully understood, nevertheless, the most important role is played by lithology, slope, drainage, landforms, absence of vegetative cover and unusual behaviour of the rainfall pattern. Sometimes, mining and seismic activity also cause increased landslide activity.

Forest fires occur quite frequently in India during summer season. A study by Forest Survey of India has revealed that

51% of the forest area in Assam and Gujarat, 93% in Arunachal Pradesh, 67% in Bihar, 69% in H. P., 46% in J & K, 45% in Karnataka, 76% in M.P., 94% in Meghalaya and Orissa, 87% in Nagaland, 58% in U.P. and 33% in West Bengal is subject to repeated annual fires. While it is not possible to assess the actual damage caused, according to the figures given in the Forest Survey of India report (1987), fire occurred as many as 17, 852 times in the country causing fire in 5,72,417 sq. km area. The total damage due to fire was over 7 million rupees during the VI Five Year Plan.

Floods in India are an annual event causing extensive damage to crops, property and life. Depending upon rainfall, snowmelt and cyclonic depressions varying magnitudes of flood occur in large parts of India. About 40 million ha area in the country is flood prone (National Flood Commission). Satellite remote sensing is capable of providing timely, comprehensive and reliable information on flooded areas, extent of damage to crops and property besides helping in flood hazard zonation.

Cyclones hit east and west coasts of the country almost every year resulting in severe loss of life, infrastructure and property. In our country cyclones normally occur during April-May (pre-monsoon cyclonic storms) and October-December (post-monsoon cyclonic storms) though they may occur during the rainy season.

Global Issues

Daunting ranges of environmental challenges are faced by the mankind resultant from human actions related to resource use and anthropogenic changes in atmosphere, land and oceans. It is now realised that many of human actions have transnational and lasting effects and affect the delicate balance in functioning of earth. Thus, scientific understanding of the earth as a combined system (biological-geophysical-atmospheric) and documenting how past human actions have modified various elements of the system is urgently needed. Global warming (as a result of increased emission of Green House Gases), ozone depletion, marine pollution, loss of biodiversity etc. are the major global environmental issues.

Information Needs and RS Compliance

Remote sensing has come a long way since 1972 when USA launched first remote sensing satellite. Presently, a number of remote sensing satellites providing data sets in different spatial and spectral resolutions are available. India launched its first remote sensing satellite IRS 1A in March 1988. Since then a galaxy of IRS series of satellites such as IRS 1B, IRS 1C, IRS 1D, IRS P3, IRS P4, IRS P6 (Resourcesat-1) and Cartosat -1 have been orbited providing regular services for monitoring earth and its environment. This has enabled monitoring of earth's environment from continental to local scales embodying different facets of environment. In general, the Remote Sensing based information is used for a) Environmental clearance of projects based on present environment status; b) Generation of environmental management plan (EMP); c) Environmental Auditing; d) Extension of ongoing projects; and e) Public litigation cases in courts as legal documents.

Environmental Assessment and Monitoring: Information Needs and RS amenability.

Ecosystem Inventory and Monitoring

Variable	Information Need	Status/Remarks
Forest Ecosystem		
Forest cover, type and other spatial information	Inventory at national/regional level	Operational
	Type	Operational
	Crown Density	Operational
	Regeneration/ afforestation/ deforestation, encroachment	Operational
	Disease infestation	Feasible
Canopy structure	Leaf area index	AWiFS/AVHRR/Vegetatio
	Canopy closure	Aerial photography
	Canopy height	Radarsat
	Health vigour	Resourcesat
Canopy Processes		
Phenology	Leaf-on->leaf off	AWiFS/AVHRR, MODIS, IRS
NPP	Net Primary Productivity	AWiFS/AVHRR, MODIS, IRS
FPAR	Fraction PAR absorbed	AWiFS/AVHRR, MODIS, IRS
Biomass	Monoculture	Feasible; better spatial and radiometric resolution needed
	Natural forests	Feasibility shown
Biodiversity	Indicators of biodiversity Species Richness	Landscape possible, 1/2 m spatial resolution, Field sampling
Wildlife/Faunal Resources	Habitat evaluation; Corridor mapping	Operational; better spatial (10m) resolution needed

Wetland Ecosystems

Spatial information	Inventory and monitoring Structural components/habitat types Aq. veg/weeds extent Aquatic vegetation type	Extent, type, First inventory carried on 1: 250 000/1: 50 000 scale Operationally possible on 1:50 000/ 1:25 000 scale Emergent, floating weeds possible to delineate on 1: 25 000 or larger scales Large patches can be discerned
Water quality	Pollution sources/waste outfalls	Large out falls possible to map operationally
	Thermal pollution	Possible to detect
	Phytoplankton/algal blooms	Inland wetlands not established
	Turbidity/sediment load	Possible in conjunction with field data
	Oil pollution	Possible in coastal waters
	Pollutant species	Ground based studies
Biodiversity	Habitat diversity Fragmentation Edge to area ratio Habitat-biotic assemblages etc.	Possible at 1: 25 000/10 000 scales
Catchment characteristics	Land cover, soils, landforms, hydrologeomorphology, etc.	Possible at 1: 25 000/10 000 scales
Action Plan for Conservation & Management	Integration and criteria based action plan generation	Possible at 1: 25 000/10 000 scales

Environmental Pollution

Land Pollution	Identification and monitoring municipal wastes, sanitary landfills, derelict land Extent of open cast mines, quarries, soil heaps, overburden Land affected by mine dust/run-off, mine fires Thermal power generation-area of influence	Mapping for large areas possible Operational for large areas Operational for large areas Feasible; limited utility.
Air Pollution	Plumes (thermal and cement plants) detection, direction and spread Impact on vegetation	Mapping possible operationally (1:50 000/1:25 000 scales) Feasible at gross level in highly polluted areas. Operational systems not available. Operational systems not available.
Water Pollution	Impact on aquatic habitat Pollutant species (SO_x, NO_x, SPM etc.)	Given under wetlands

Environmental Hazards

Landslides	Detection, monitoring and hazard zonation	Possible
Forest Fires	Fire detection (active)	Possible (NOAA/IRS AWiFS, MODIS data)
	Fire scars	Possible to do operationally- post-event
	Fire prone area mapping	Can be operationally done
Mine Fires	Detection and monitoring	Possible with thermal data
Floods	Damage assessment	Possible to do operationally
Cyclones	Detection, prediction	Possible to do operationally
River Pollution	Detection	Large point sources possible (coloured)

Mapping and Monitoring - Considerations

Remote sensing technology in the visible, infrared and microwave regions of electromagnetic spectrum has been used in a number of studies for environmental assessment and monitoring. In the case of analysis of temporal digital data from the same or different satellites appropriate corrections should be applied for changes in sun angle, sensor parameters (gain, bias, etc.) and the shift in the reflectance caused due to atmospheric scattering. The multi-date data should be accurately registered either with respect to each other or to the available topographical maps. The main considerations in the selection of data are:

- Ideally, the same date/season data should be used for detecting ecological changes. However, more than one date data should be used for mutually distinguishing cover types. In case of hilly areas, data of higher sun angle should normally be selected to avoid the problem of shadows.
- In the case of analysis of temporal digital data from the same or different satellites appropriate corrections should be applied for changes in sun angle, sensor parameters (gain, bias, etc.) and the shift in the reflectance caused due to atmospheric scattering.
- The multi-date data should be accurately registered either with respect to each other or to the available topographical maps.

Visual Analysis

In this method, the multi-date data are separately analysed and change maps generated subsequently for monitoring. The salient steps of the visual analysis technique are:

Preparation of base maps: Scale is decided by the aims and objectives of the study and type of satellite data. Normally, 1:250,000, 1:50,000 and 1:25,000 scale base maps are prepared for various ecosystem types. Prominent land features like rivers, roads, railways and major settlements are marked on the base maps which also serve as controls while doing interpretation/ mapping.

Preliminary visual analysis: With the help of elements of image analysis (tone, texture, size, shape and association etc.) and ancillary information, like available maps, preliminary thematic maps on the desired scale are prepared. In case of maps having spatial domain, minimum mapping size (normally 3 x 3 mm) is taken into consideration while delineating classes.

Ground truth data collection: For preparing various thematic maps and ecological parameters ground truth data specific to the theme under consideration is collected and correlated with the image characteristics. Thus, an interpretation key is prepared.

Visual analysis of current and archived data: Visual analysis is carried out for the current data. To prepare maps for the earlier dates, the template of the map of current data is made and the changes incorporated using optical enlarger equipment for preparing the maps for the past years for which no ground truth exists.

Preparation of 'change' maps: From the set of the maps prepared for various years, change map is prepared by delineating those areas, which have undergone changes.

Digital Analysis

The digital analysis techniques for environmental assessment and monitoring can be broadly grouped into the following:

- Image differencing
- Orthogonal transformations
- Image ratios
- Classification comparison

Image Differencing: The grey values (DN values) of registered multiband images of two date data are subtracted on a pixel-by-pixel basis to obtain difference image in each band. By convention, the previous date image (reference image) is subtracted from the later date image of the same area. The resultant image shows 'positive' and 'negative' digital change values.

The operation can be expressed as: $D_b = A + (X_{2b} - X_{1b}) * B$

Where, D_b is difference image in band b, X_{1b} and X_{2b} are the digital values of two date images for band b, and A and B are the constants for stretching the output values to the display range.

The resultant difference image in each band is, approximately gaussian in nature with no-change pixels centred around the mean, while the tail contains information about the changed areas as a function of time in a given area suffering from the problem of land pollution. A critical element of the image differencing method is in deciding where to place the threshold boundaries between change and no-change pixels. A change threshold has to be specified by the user. If a single type of change is to be detected, say from category A to B, the threshold may be specified as the difference in the reflectance between A and B. If the surface area, contains several categories of change, statistical threshold at 1, 2 or 3 Standard Deviation can be applied. The threshold can also be decided interactively by adjusting it for a training area and then using it for the entire image.

Orthogonal transformations: These methods of change detection involve the transformation of the data set to a new set of axes, which are orthogonal to each other. The two most commonly used methods are principal component analysis (PCA) and Kauth-Thomas (KT) transformation. PCA transforms the axes of the raw multispectral data into a set of axes orthogonal to each other and ordered in terms of variance. The PCs are obtained from the Eigen vectors of the variance-covariance matrix or correlation matrix. The first component (PC1) generally accounts for about 90-95 per cent of variance. Change detection using PCA involves comparison of the PCs of individual dates by image differencing, classification etc. In

Kauth-Thomas (KT) Transformation remotely sensed digital data is combined linearly so that the transformed bands give information on specific characteristics such as soil and vegetation. Originally designed for crop studies using LANDSAT MSS data, the transformation generates a set of axes representing soil brightness, vegetation greenness, yellow stuff but not so in case of IRS data. Analysis for change detection can be done either by image differencing or superposing one date transformation over the other.

Image Ratioing: Band ratios are useful in separating physical classes of land cover. A ratio image of the scene effectively compensates for the brightness variation. Simple band ratio (IR/R) or other variations such as NDVI (normalised difference vegetation index) (IR-R)/IR+R) of the two dates have been used in a number of studies to monitor ecological changes either by image differencing or superposition of multi-date images.

Another variation of the ratio method is to divide the digital values of each pixel for one band of later date by the corresponding one of the earlier date. Unchanged features give a ratio value of 1, while changed features yield ratios less than or greater than 1. These values are stretched to the display range. The resultant image then can be subjected to thresholding or a FCC of the ratio image can be made to bring out the changes.

Classification Comparison: This is one of the most commonly used methods for comparing the ecological changes. In this method two scenes are independently classified and then compared for possible changes. Alternatively, classified images can be superposed or differenced to get the changes. One of the major drawbacks, however, is the accuracy of the results of classification. When multi-date data is considered together, accuracy is the product of individual accuracies. Classification can also be done on raw data or on higher components of PCA or on ratio images etc.

Selected Examples

Forest Ecosystems

Information about the extent of forest areas and the growing stocks are the most important components of inventory and monitoring. Remote sensing has emerged as the most important tool for inventory, monitoring and management of forest resources at local, regional and continental scales. A critical component of forest inventory and monitoring is the scale of the mapping and selection of proper season data. Many of the parameters are needed on a stand or an individual tree basis. Optical and microwave sensors do not measure forest parameters directly, but rather are constrained by physical principles applied to a complex physical environment in developing relationships between backscattered microwave or optical radiation and the desired forest parameters.

Forest Survey of India is biennially monitoring Forest cover at the national level using satellite remote sensing data since 1985-87. According to 2000 forest cover assessment 19.39 percent of the geographical area in the country is under forests, though dense forest is only 11.48 percent. Studies carried out so far have shown the potential of satellite data for forest type, density, volume estimation of timer trees, biomass and productivity at various levels. Remote sensing and GIS have also found a significant use in the preparation of Working Plans for the management of forest resources.

Protected Areas

Protected area management is of paramount significance for conservation of biodiversity and maintaining required area under wilderness. India has designated about 4.5 per cent geographical area under biosphere reserves, national parks and sanctuaries for *in situ* conservation of flora and fauna. 12 Biosphere Reserves have been set up so far to conserve and protect unique and representative ecosystems in the country. These are (Nigeria, Nanda Devi, Nortek, Mamas, Sunder bans, Gulf of Manner, Great Microbar, Similar, Dibru-Saikhowa, Dehang Debang, Pachmarhi, Kanchanjanga).

Protected area management presupposes availability of accurate information on food, water and shelter. With the availability of high-resolution satellite data (IRS 1C/1D, IKONOS), GIS and ground sampling wildlife habitats can be evaluated more accurately for management of protected areas.

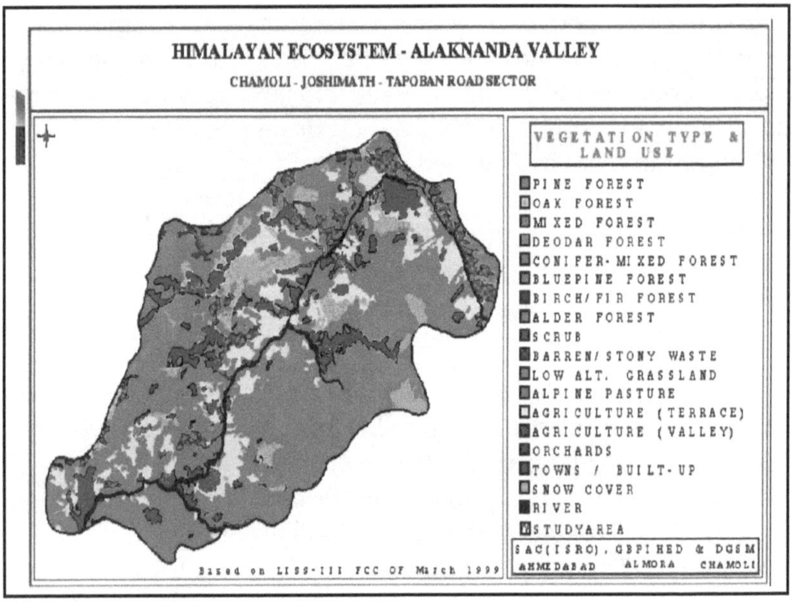

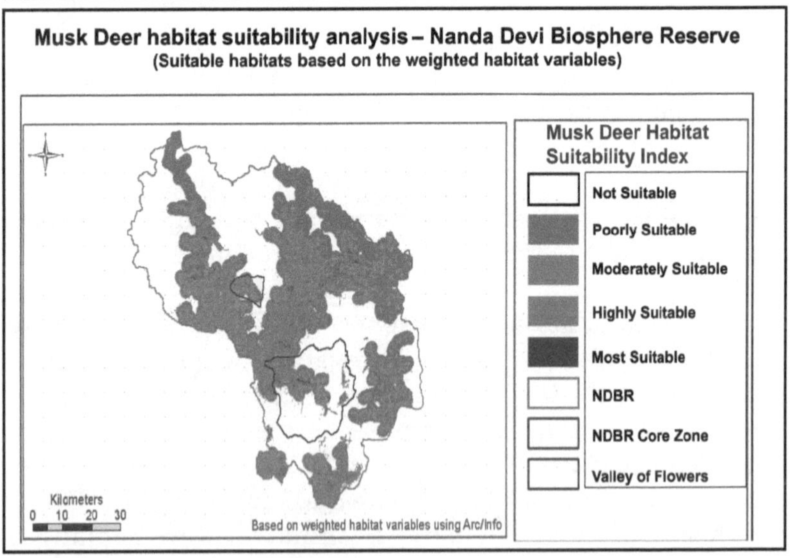

Wetland Ecosystems

Wetlands, the cradles of biodiversity, are among the most productive ecosystems housing more or less all floral and faunal taxonomic units. They are not only important for the survival of fishes, birds, turtles and crocodiles etc., but also are habitats of many rare and endangered species, as for example, the Brown Antlered Deer (*Cervus eldi eldi*) in the swamps of Loktak Lake in India. Migration of birds is another reason for faunal diversity in wetlands. Due to reckless filling and improper use of wetlands a large number of them are either lost or are incapable of fulfilling their designated functions/uses. Wetlands are under the constant threat resultant from encroachment, siltation, aquaculture development, pollution and taming of rivers etc. The cumulative strains on wetlands is increasingly becoming evident in the form of a) Decrease in biological diversity, particularly endemic and endangered species; b) Deterioration of water quality; c) Sedimentation and shrinkage in areas under wetlands; d) Reduction in migratory bird population, and the productivity of fish and other fauna; e) Prolific growth of unwanted aquatic biota (flora and fauna).

The 1971 Ramsar Convention on Wetlands of International Importance especially as Waterfowl Habitat is the oldest conservation convention. It owes its name to its place of adoption in Iran. It came into being due to serious decline in populations of waterfowl (mainly ducks) and conservation of habitats of migratory waterfowl. Convention provides framework for the conservation and 'wise use' of wetlands. Ramsar Convention is the first modern global intergovernmental treaty on conservation and wise use of any ecosystem type. Ramsar convention entered into force in 1975 and contracting parties from all over the world. As of October 2003 there are 138 contracting countries, which have identified 1313 wetlands as Ramsar sites. These sites cover an area of 111 million ha. India has declared Keoladev National Park, Loktak, Chilka, Sambhar, Wular, Harike, Ashtamudi, Bhitarikanika, Bhoj, Deepor beel, East Calcutta wetlands, Kolleru, Point Calimere, Pong dam lake, Sasthamkotta, Vembanad-Kol, Tso Moriri, Kanjali, and Ropar as Ramsar sites.

Wetland definitions

Scientists working in their specialized fields such as botany or hydrology have defined wetlands according to their perception. A botanical definition focuses on the plants adapted to flooding and/or saturated soil conditions, while hydrologists emphasize the position of water table relative to the ground surface over time. Generally, two types of wetland definitions are in vogue, one, which are broad, and the other, which are narrow. Narrow definitions look at wetlands as ecotones, i.e. transition zone between terrestrial and aquatic communities (Cowardin *et al,* 1979) while broad definitions categorise wetlands keeping in view management perspectives (Ramsar Convention Website).

Ramsar Convention

"Submerged or water saturated lands, both natural and man made, permanent or temporary, with water that is static or flowing, fresh, brackish or salt, including areas of marine water the depth of which at low tide does not exceed 6 meters".

US Fish and Wildlife Service

"Wetlands are lands transitional between terrestrial and aquatic systems where the water table is usually at or near the surface or the land is covered by shallow water level. For purposes of this classification, wetlands must have one or more of the following three attributes: 1) at least periodically, the land supports predominantly hydrophytes; 2) the substrate is predominantly undrained hydric soil; and 3) the substrate is nonsoil and is saturated with water or covered by shallow water level at some time during the growing season of each year".

Inventory and monitoring

Conservation of wetlands has been neglected not only in India but also through out the world. This is primarily due to lack of scientific inventory. In a landmark initiative, the Ministry of Environment and Forests Govt. of India sponsored a project to Space Applications Centre (ISRO), Ahmedabad for inventory

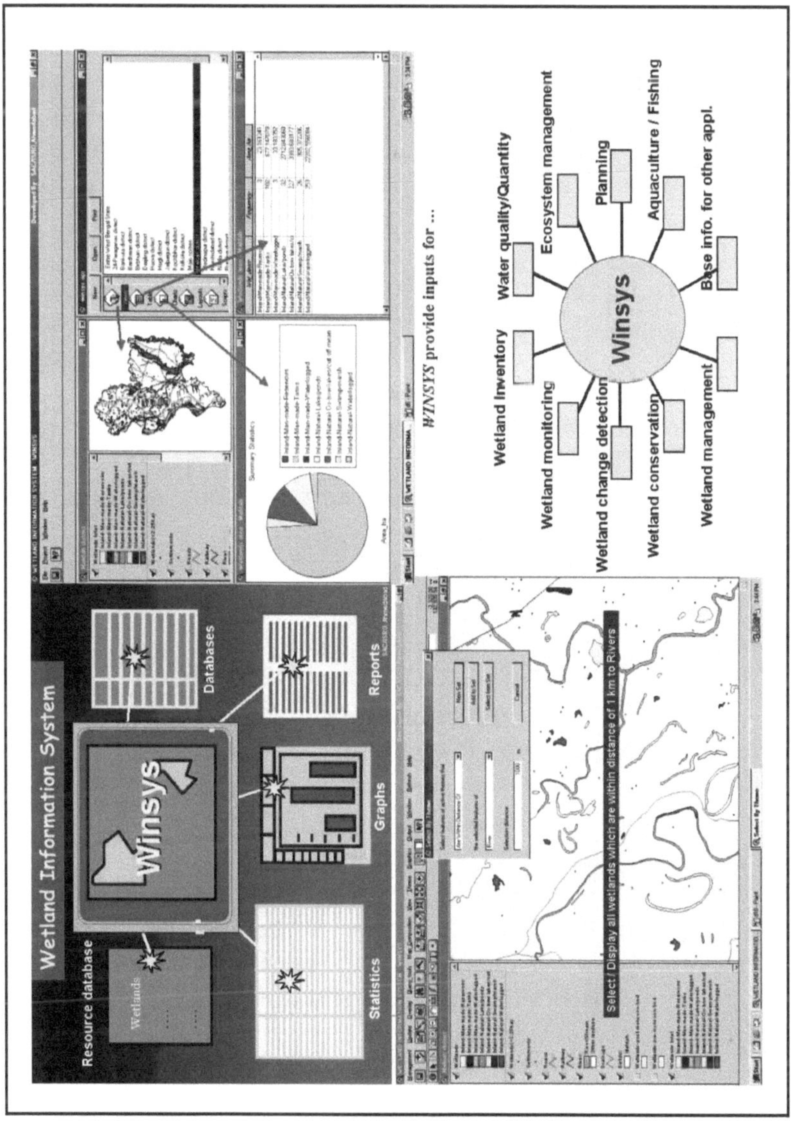
Wetland Information System
Resource database
Wetlands
Winsys
Statistics
Graphs
Databases
Reports
Select / Display all wetlands which are within distance of 1 km to Rivers
WINSYS provide inputs for ...
Wetland Inventory
Water quality/Quantity
Wetland monitoring
Ecosystem management
Wetland change detection
Planning
Wetland conservation
Aquaculture / Fishing
Wetland management
Base info. for other appl.
Winsys

of wetlands in the country using Indian Remote sensing Satellite (IRS) data. Mapping was done on 1:250 000 scale for most of the states and all wetlands greater than 56.25 ha were delineated. For North-Eastern states, Union Territories and Punjab, Haryana, Goa, Himachal Pradesh, Sikkim and West Bengal work was carried out on 1:50 000 scale and all wetlands units bigger than 2.25 ha were delineated. A total of 1576 maps have been prepared for the entire country. There are 27403 wetland units in the country bigger than the minimum mapping unit (56.25 ha or 2.25 ha depending on the scale) occupying 7581871 ha area. In addition, 80, 000 wetlands which are smaller than 56.25 or 2.25 ha depending on the scale of mapping) also exists. An important feature of the study is the creation of attribute database of the wetlands in the country. The database includes information on SOI (Survey of India) topographical map number, state, district, tehsil/taluka, nearest village, central co-ordinates, wetland type (24 classes), wetland area, water spread as manifested on post and pre-monsoon season imagery, turbidity, aquatic or other vegetation within the wetlands etc. State-wise atlases have been also prepared for a ready reference.

National Wetland Information System

For effective monitoring and management of wetland resources a GIS based Wetland Information System is the prerequisite. In order to make an easy use and quick access of the voluminous data generated through analysis of remote sensing a Wetland Information System (WINSYS) has been designed and developed (Patel *et al,* 2003). The easy to use 'Graphic User Interface (GUI)' of *Arcview GIS* has been used for organizing spatial database. Using the Winsys one can query for the state, district or type etc. The information system has been developed for the state of W. Bengal with a provision for incorporating all the states and individual wetlands.

Indian Wetland Classification System

Wetlands exhibit enormous diversity according to their genesis, geographical location, water regime and chemistry, dominant plants and soil or sediment characteristics. Existing definitions of wetlands recognize that wetlands are driven by wetland hydrology (permanent or periodic inundation) hydric soils, and characteristic hydrophytes. Wetlands are normally associated with the following major systems:

Marine (coastal wetlands including rock shores and coral reefs).

Estuarine (including deltas, tidal marshes and mangrove swamps).

Lacustarine (wetlands associated with lakes)

Riverine (wetlands along rivers and streams)

Palustarine (meaning 'marshy' - marshes, swamps and bogs)

In addition, man-made wetlands such as fish and shrimp ponds, saltpans, reservoirs, gravel pits, sewage farms and canals etc. are also considered as wetlands as per Ramsar definition. A National Wetland Classification system was evolved after considerable deliberations with the experts and planners for carrying out national wetland inventory. The classification system besides including all wetlands incorporates deep water habitats and impoundments. Main criteria followed in this system are: i) Wetland hydrology, *i.e.* manifestation of water on the satellite imagery; ii) Wetland vegetation – mainly hydrophytes and other aquatic vegetation in a part or whole of the water body as observed on satellite data.

Inland Wetlands	1. Natural	1.1 Lakes/Ponds 1.2 Ox-Bow Lakes/Cut-Off Meanders 1.3 Waterlogged (Seasonal) 1.4 Playas 1.5 Swamp/Marsh
	2. Man made	2.1 Reservoirs 2.2 Tanks 2.3 Waterlogged 2.4 Abandoned Quarries 2.5 Ash Pond/Cooling Pond
Coastal Wetlands	3. Natural	3.1 Estuary 3.2 Lagoon 3.3 Creek 3.4 Backwater (Kayal) 3.5 Bay 3.6 Tidal Flat/Mud Flat

Contd...

Contd...

		3.7 Sand/Beach/Spit/Bar 3.8 Coral Reefs 3.9 Rocky Coast 3.10 Mangroves 3.11 Salt Marsh/Vegetation 3.12 Other Vegetation
	4. Man made	4.1 Salt Pans 4.2 Aquaculture Ponds

- *Rivers have not been included in the classification system, however, are shown in the maps.*
- *The extent of vegetation, if present, in the inland wetlands is indicated on the maps.*
- *Qualitative turbidity ratings (low, moderate, high) given for inland wetlands, wherever possible.*
- *Wetlands put to regular agriculture use have not been included.*

Wetland Specific Resources Information System

For effective conservation of important wetlands such as Ramsar sites, notified wetlands and bird sanctuaries etc. it is prudent to develop a GIS based information system incorporating remote sensing derived information on wetland *per se*, catchment characteristics, biophysical data on water quality, flora and fauna, and socio-economic set up etc. in a GIS environment for easy retrieval and updating, generating action plans and monitoring the success. In India, Space Applications Centre, Ahmedabad, has developed such a system for Loktak Lake.

Loktak Lake Resources Information System (LRIS)

Loktak Lake, located between longitudes 93^0 46′ & 93^0 55′ E and latitudes 24^0 25′ & 24^0 42′ N in the State of Manipur is a shallow wetland, with varying depth of 0.5 m to 1.5 m. and is considered as the lifeline of the people of Manipur due to its importance in their socio-economic and cultural life. It is the largest natural freshwater lake in the northeastern region and plays an important role in the ecological and economic security. A large population living around the lake depends upon the lake resources for their sustenance. Loktak Lake has rich biodiversity and has been designated as a Wetland of International Importance under.

Type-wise estimates of Wetlands in India

Wetland Category	No	Area (ha)
Inland Wetlands - Natural		
Lakes/Ponds	4646	679530
Ox-Bow Lakes	3197	151051
Waterlogged (Seasonal)	4921	285744
Playas	79	118519
Swamp/Marsh	1814	197784
Total	14657	1432627
Inland Wetlands - Man-Made		
Reservoirs	2208	1481987
Tanks	5549	558344
Waterlogged	892	77302
Abandoned Quarries (Water)	105	5774
Ash Pond/Cooling Pond	33	2881
Total	8787	2126288
Total Inland	23444	3558915
Coastal Wetlands- Natural		
Estuaries	97	153966
Lagoons	34	156403
Creeks	241	19230
Backwater	32	17075
Tidal /Mud Flat	663	2362056
Sand/Beach/Spit/Bar	772	421019
Coral Reefs	487	84137
Rocky Coast	85	17686
Mangroves	858	340055
Salt Marsh/ Marsh Vegetation	161	169840
Other Vegetation	117	139102
Total	3497	3880569
Coastal-Man-Made		
Salt Pans	106	65496
Aquaculture Ponds	356	76891
Total	462	142387
Total Coastal	3959	4022956
Total India	27403	7581871

Note: In addition, 80,000 wetlands smaller than the minimum-mapping unit were also detected and shown on the maps. Of these 10630 are smaller than 56.25 ha (1:250 000 scale) and 69370 less than 2.25 ha (1:50 000 scale, majority being in W. Bengal). Assuming each wetland occupies 50 % area of the minimum mapping unit; these wetlands might occupy 675979 ha area. Thus the total wetland area excluding rivers, canals and rice in the country using Remote sensing data is estimated as 8257850 ha (~8.26 Mha). Rice occupies 42.2 Mha area (Manorama Year Book). Five river systems in the country have 28000 km of length; Canals and irrigation channels are 113000 km in length.Source: Garg et al, 1998

State-wise Distribution of Wetlands in India

State	G. Area (km²)	Wetland Area (ha)		Total Wetland Area (ha)
		Inland	**Coastal**	
1: 250 000 Scale				
Andhra P.	275075	366609	185454	552063
Bihar	173877	177683	-	177683
Gujarat	195980	209206	2508263	2717469
Jammu & Kashmir	222236	406780	-	406780
Karnataka	192204	254015	17826	271841
Kerala	38863	34200	93731	127931
Madhya P.	443446	294118	-	294118
Maharashtra	307609	284942	100030	384972
Orissa	105707	162774	185432	348206
Rajasthan	343214	344964	-	344964
Tamil Nadu	130069	161521	398705	560226
Uttar Pradesh	29411	328690	-	328690
1: 50 000 scale				
Pondicherry	492	59	833	892
Arunachal Pradesh	83740	56325	-	56325
Assam	78438	101232	-	101232
Delhi	1483	4717	-	4717
Goa	3702	2145	8736	10881
Haryana	44212	27057	-	27057
Himachal Pradesh	55673	54766	-	54766
Manipur	22327	52959	-	52959
Meghalaya	22429	2222	-	2222
Mizoram	21081	152	-	152
Nagaland	16579	918	-	918
Punjab & Chandigarh	50362	71879	-	71879
Sikkim	7096	1985	-	1985
Tripura	10486	9896	-	9896
W. Bengal	88752	143859	360447	504306
A& N Islands	8293	3204	107759	110963
D. D. & N. Haveli (incl. Daman & Diu)	498	38	1474	1512
Lakshadweep Is.	32	-	54266	54266
Total		3558915	4022956	7581871

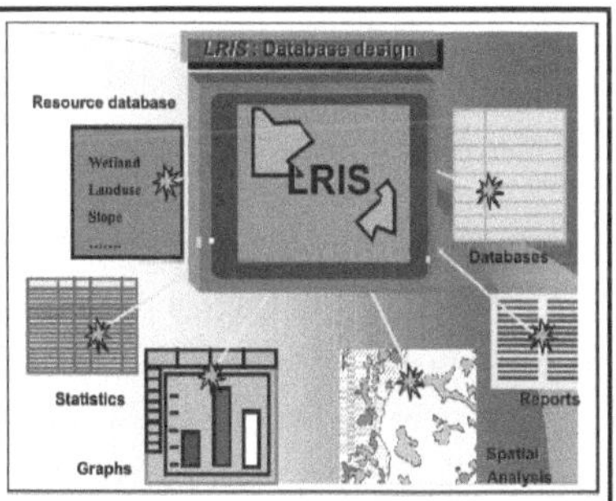

Loktak Lake during 1988, 1999 and 2002

Ramsar Convention in 1990. The Keibul Lamjao National Park, in the southern part of the lake, is a unique floating wildlife reserve and is the refuge of the highly endangered brow-antlered deer locally called '*sangai*' (*Cervus eldi eldi*). It is the only floating wildlife sanctuary in the world. It is composed of a continuous mass of floating aquatic vegetation '*phumdi*' occupying an area of 40 sq. km. The lake is an ideal habitat and breeding ground for a number of fishes and supports a significant population of migratory and resident waterfowl. The extent of water spread depends on reservoir/dam height constructed at Ithai. Aquatic plants both floating and submerged which accounts for 16773 ha (about 65 % of the lake).

Loktak Lake is deteriorating at an alarming rate due to anthropogenic activities. Deforestation and shifting cultivation (jhuming) in the catchment has accelerated the process of soil erosion resulting in lake shrinkage. The problem has further been aggravated due to prolific growth of floating weed-mats locally called as '*phumdis*'. Besides, the nutrients from the catchment area, domestic sewage from Imphal city are carried by the Nambul River, which finally discharges into the lake. Pesticides used in the agricultural fields are also washed off into the lake. In addition to above threats, the encroachments in the lake by construction of fishponds, roads and settlements have gradually led to the shrinkage of the lake area.

Wetland Biodiversity Conservation

Remote sensing, GIS and GPS have started playing a major role in wetland biodiversity conservation. Space Applications Centre (ISRO), Ahmedabad has initiated a project for Wetland Biodiversity Conservation in Ramsar sites using remote sensing, GIS, GPS and field sampling. Satellite data is being analyzed using visual and/or digital analysis techniques (Mapping scale: 1:25000 scale or larger) for eliciting temporal structural changes in the wetlands as well as catchments. Taxonomic information on flora and fauna for the assessment of biodiversity will be collected using field surveys. Indices will be developed based on each taxonomic unit (family/order). Factors and constraints related to biodiversity as manifested from analysis of satellite data will be identified. Biophysical and ecological data will be gathered through field surveys. Multimetric indices will be developed for subsequent monitoring of biodiversity. Catchment characteristics having direct bearing on the wetland will be studied and subsequently integration will be done in GIS to formulate strategies for arriving at sustainable solutions for conservation of biodiversity in Ramsar Site.

Coral Reefs and Mangrove Ecosystems

Coral reefs a source of high quality protein and provide protection to the fragile shoreline from erosion. They are also hotspots of marine biodiversity. Space Applications Centre, Ahmedabad has carried out a survey of coral reefs along the Indian coast using IRS LISS II data on 1: 50 000 scale. Area under reefs in the gulf of Kachchh is about 150 sq. km, Lakshdweep 136.5 sq. km, Tamil Nadu coast 94 sq. km and A & N islands 960 sq. km. Mangroves ecosystems are an important element of coasts dotting and gregariously growing along many tropical seas. Information on extent of mangroves and dominant community types on 1: 50 000 scale for the entire Indian coast including the islands.

Methane emission from wetlands

Prediction of future climatic changes and global productivity presupposes realistic projections of carbon containing gases such as carbon dioxide and methane in the atmosphere. In order to estimate atmospheric concentration of carbon dioxide and methane contribution from terrestrial and aquatic (freshwater as well as oceanic waters) ecosystems with reasonable accuracy understanding of the relationship between the physical environment, anthropogenic activities and biological activity is of utmost significance. In case of wetlands, it requires knowledge of changes in wetland extent, and conditions and their causes (forcings), response of wetland ecosystems to global environmental change (Response), consequences of wetland extent and condition change for ecosystem sustainability and economic productivity

(Consequences), and modeling of carbon containing gases in the biosphere (Prediction). Due to its Global Warming Potential (GWP), which is 21 times higher than carbon dioxide and with a greater volume of emissions than any other gas except CO_2, methane is considered as the most significant green house gas among the GHGs. According to IPCC (International Panel on Climate Change, 1998) since 1750 atmospheric concentrations of methane have increased by 150 percent in 1998 from approximately 700 to 1,745 parts per billion by volume (ppbv).

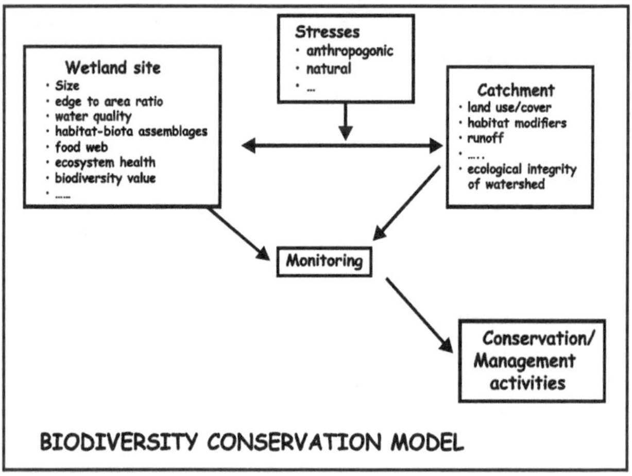

A Framework for wetland biodiversity conservation

India has been divided into 10 different biogeographic zones and 26 biotic provinces (WII, 2000). Space Applications Centre (ISRO) has estimated first country level methane emissions from wetlands to be 1.27 Tg/CH4/year. Annual methane emission in term of methane emissions per sq. km area is has been found to be 388 kg/sq. km. Ecozone-wise methane estimates from wetlands are given on page 29.

The first major study concerning use of remote sensing data for EIA studies in India began in 1986 when the project on 'Impact of mining activities and super thermal power stations on environment' was undertaken by Space Applications Centre, Ahmedabad. The project was taken up with the financial assistance from the Ministry of Environment and

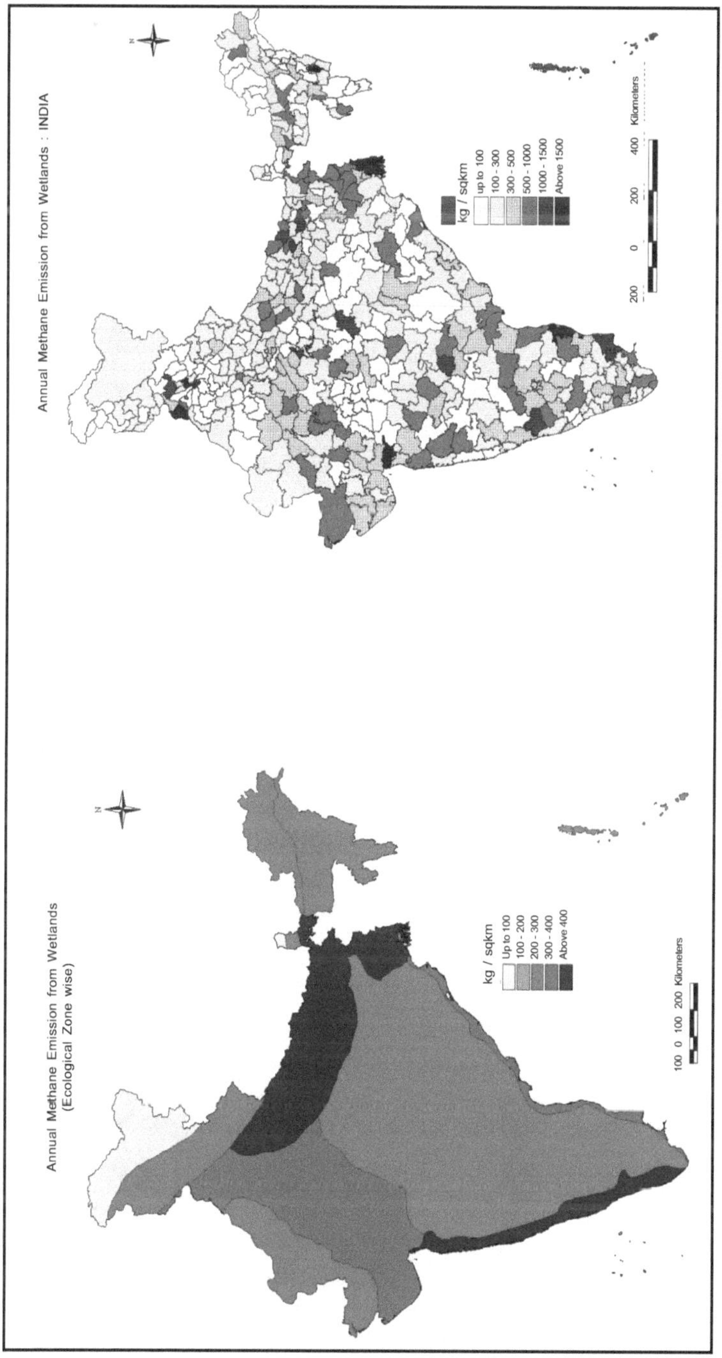
Annual Methane Emission from Wetlands : INDIA
kg / sqkm
up to 100
100 - 300
300 - 500
500 - 1000
1000 - 1500
Above 1500
200 0 200 400 Kilometers
Annual Methane Emission from Wetlands
(Ecological Zone wise)
kg / sqkm
Up to 100
100 - 200
200 - 300
300 - 400
Above 400
100 0 100 200 Kilometers

Forests, Govt. of India under the auspices of SC-B. This was a landmark study in which methodologies were developed for 'digital change detection', detection of coloured water pollution and impact of air pollution (fly ash) using remote sensing techniques for regional EIA. This was followed by environmental impact studies of Tehri dam in Uttar Pradesh and Narmada dam in Gujarat/Maharashtra.

Mining

A number of studies have been carried out by SAC to assess the impact of mining activities on many facets of environment. Studies in Korba, Chattisgarh have been carried out using multitemporal satellite data to monitor changes in land cover and the areas affected by coal mining and flyash deposition. Mining activities have affected an area of 64.30 sq. km especially due to coal dust deposition in the environs of coalmines and dust migration through drainage. Similar studies have also been carried out for Talcher where mining activities started in 1921 at Hardidhua. In the study for Dehradun-Mussoorie limestone belt active and abandoned mines have been mapped using Landsat MSS, Landsat TM and IRS LISS II data of 1972, 1985 and 1988. Active mine area which was 1.5 sq. km in 1972 has reduced to 0.6 sq. km in 1988 due to Supreme Court verdict of 1985. Consequently in 1988 abandoned mines occupied an area of 1.4 sq. km. Due to mining activities, the river beds of some streams, notably Baldi and Nun have also either widened or changed.

Thermal Power Generation

Remotely sensed data has been used to study the changes in land cover and the deposition of fly ash in Korba and Talcher. In Korba, by using IRS LISS II data, it has been brought out that about 605 sq. km area has been affected by flyash deposition out of which 131 sq. km is severely affected while 474 sq. km is moderately affected as of 1989. In Talcher, the total area affected by fly ash deposition has increased from 25 sq. km in 1979 to 117 sq. km in 1987. However, the impacted areas could not be delineated on IRS LISS II imagery of October

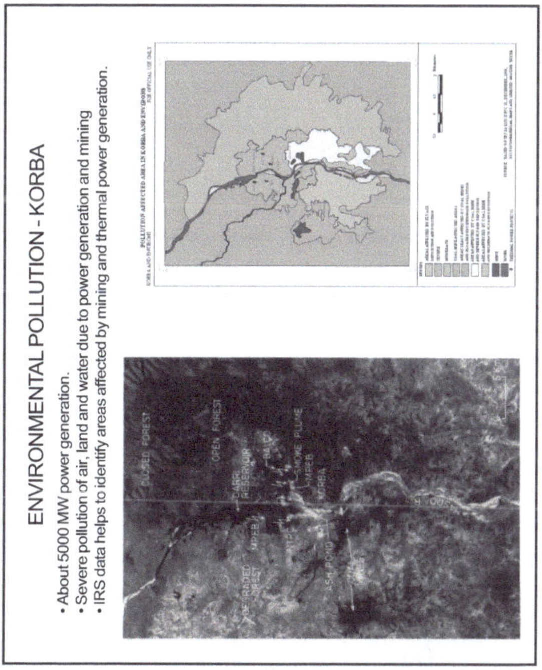
ENVIRONMENTAL POLLUTION - KORBA
• About 5000 MW power generation.
• Severe pollution of air, land and water due to power generation and mining
• IRS data helps to identify areas affected by mining and thermal power generation.
CLOSED FOREST
OPEN FOREST
DEGRADED FOREST
MPEB
BALCO
SMOKE PLUME
KORBA
ASH POND

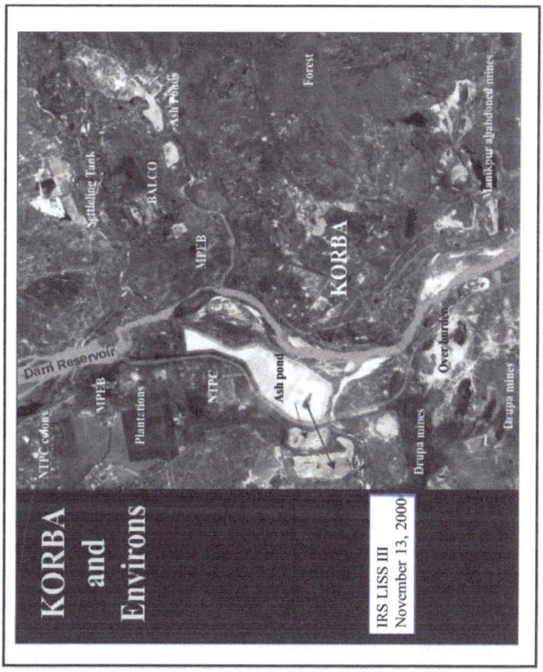
KORBA
and
Environs
IRS LISS III
November 13, 2000
NTPC colony
MPEB
Plantations
NTPC
Dam Reservoir
Ash pond
Settleling Tank
BALCO
Ash Ponds
MPEB
KORBA
Forest
Over burden
Drupa mines
Drupa mines
Manikpur abandoned mines

1988 and also subsequent years because of the commissioning of ESP.

Prevention and Control of Pollution

Prevention and control of pollution (air and water) requires time series data on various pollutant species in time and space. However, sensors capable of sensing horizontal and vertical distribution of pollutants at orbital altitudes are presently not available and potential of currently orbiting satellites in this regard is limited. Nevertheless, they do provide certain critical information useful in prevention and control of pollution.

Air Pollution: Commercial systems are available now for monitoring ambient air quality, which can be mounted on aircraft, but space sensing of air pollutants is still in infancy. Sensors based on pulsed fluorescence analysis, absorption of scattered solar UV radiation, gas match filter radiometry, non-dispersive principal, lasers etc. can be mounted in an aircraft for regional and point monitoring of SO_2, NO_2, O_3, CH_4 etc. with reasonably fair accuracy. Currently orbiting satellites, however, can detect smoke plumes from large industrial installations or areas.

Water Quality: Water quality studies have conventionally being carried out using *in situ* measurement of various pollutants. Remote sensing for water pollution started with the measurement of spectral response of water bodies artificially polluted under controlled conditions and their subsequent correlation with radiance/reflectance values. However, the studies could not be extended to orbital altitudes due to non-availability of specific sensors onboard various remote sensing satellites.

Environmental Hazards

Environmental hazard is the occurrence at a given site or within a given geographic area and within a given interval of time of a physical phenomenon capable of causing a great loss of value. Disasters result from the impact of the hazard on man-made structures, components and systems that are weak and incapable of resisting the forces unleashed. Hazards such as cyclones, floods, earthquakes, landslides and forest fires etc. are many times associated with loss of life and property and bring untold miseries to the people. Many of these hazards are caused or accentuated by human activities like deforestation, unplanned road construction and tourism etc. Despite the fact that huge amounts (of the order of billions) are spent every year to minimise the damage and mitigate the hazards, their impacts are on the rise. Environmental hazards studies lay emphasis on development of measures for the assessment, prediction and prevention. Satellite remote sensing has proved its utility in addressing many environmental issues resultant from many of these hazards.

Landslides

The Himalayas are the youngest and mightiest of all the mountain systems in the world extending for about 2500 km from east to west and containing boundaries of India, China, Pakistan, Tajikistan, Afghanistan, Nepal and Bhutan. Remote sensing in conjunction with other collateral data has proved very useful in landslide detection and hazard zonation.

Landslides in the Alaknanda Valley

Landslide occurrences in the Alaknanda valley are very common and have been resulting in traffic disruption, loss of life and property over the years. In the recent times, however, the frequency and severity of landslides and slope instability have increased manifold causing wide concern. August 1998 Okhimath landslide surpassed all the previous records of landslide calamities in this region resulting in loss of life and property in the Madhyamaheswar and the Kaliganga valleys. In fact, the entire Bhenti village was wiped out as a result of this tragedy. Landslides in the Alaknanda valley are the consequence of natural as well as human activities. Geologically, the area falls in the Lesser and Higher Himalayan Zones. The Lesser Himalayan Zone is overlain by Central Crystallines and separated by a major structure known as Main Central Thrust (MCT). The major rock types of the Lesser Himalayan Zone are quartzite, metabasics (Vaikrita Group), granites/gneisses, dolostones etc. Road building activity and rainfall act as the triggering factors and result in road blockade and frequent damming of rivers. Thematic maps on lithology, land use, slope & aspect, geomorphology, geological structures, drainage and road buffers were prepared and subsequently integrated in GIS environment (Schematic).

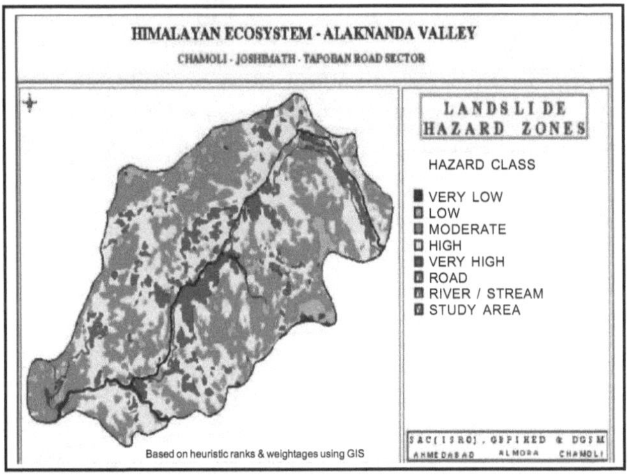

Landslide Hazard Zonation

Landslide Hazard Zone Map has been prepared for the entire Alaknanda valley in GIS by integrating various theme layers. A representative map of Chamoli-Joshimath-Tapovan Road sector is shown in the following figure:

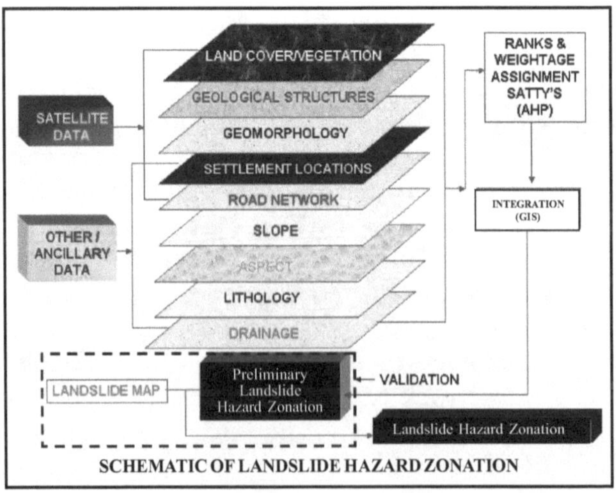

SCHEMATIC OF LANDSLIDE HAZARD ZONATION

Forest Fire

Forests are often at risk of being destroyed by forest fires. Remote sensing can be used in efforts to reduce the risk and minimize damage if a fire occurs. Weather information, such as measurements of precipitation and temperature, allows foresters to calculate risk assessments and isolate the areas most susceptible to fire. Those areas can be closely monitored by satellites, such as IRS AWiFS, MODIS, high resolution Advanced Very High Resolution Radiometer (AVHRR), Landsat TM and SPOT. Images from these satellites are readily available and small fires show up on them almost immediately. Remote sensing contributes to fire-fighting efforts, as well. Data on wind direction and speed, and the dryness of surrounding areas can help predict the directions and speed at which a fire spreads. With this information, fire fighters can be dispatched with maximum effectiveness and safety, and fires can be put out before they cause much damage.

Radar and thermal sensing allow for constant observation of fires, unaffected by clouds, smoke, or other conditions that hinder aerial observation. Currently orbiting satellites provide information on monitoring of fire scars or burned area

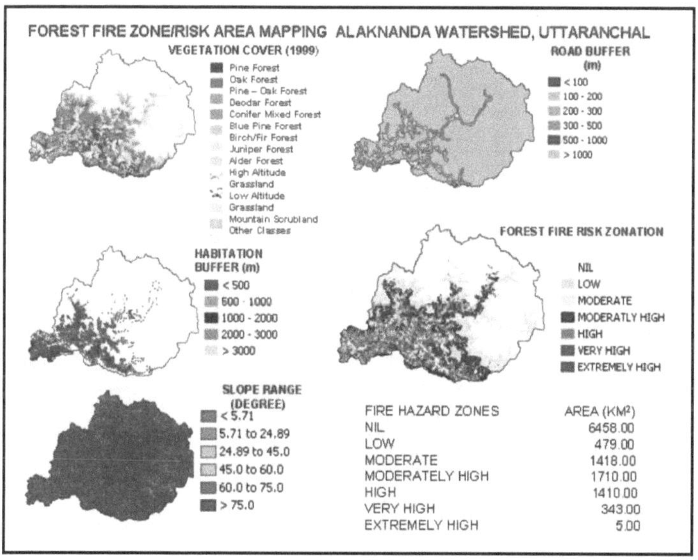

assessment, fire biomass characteristics, fire prone area mapping and fire potential. Methodology for Fire Prone Area mapping are developed on 1:50 000/25 000 scales.

Operational fire monitoring, assessment (real time) and management system comprising satellite remote sensing and communication system is to be developed. Fire prone area/ hazard mapping at national level is to be carried out as part of preparedness. 3-5 μm data will provide information on fire location, fire line movement and intensity with better accuracy. High resolution geosynchronous satellite data for continuous monitoring of forest fires, assessment of damage due to forest fire will be possible.

Earthquakes

Earthquakes are known to cause extensive damage and are among the most unpredictable of natural disasters. India has a long recorded history of earthquake devastation, the recent being in Uttarkashi (October 20, 1991), Killari (September 30, 1993) and Chamoli (March 29, 1999) Remote sensing data has been found quite useful in assessing the surficial changes subsequent to the earthquake and damage assessment. RS data is also been useful in morphostructural zoning. Many organisations in the country (GSI, WIHG, NGRI, IIT-Roorkee, SAC, NRSA and IIRS) are involved in the work related to earthquakes. IMD (India Meteorological Department) is the agency responsible for collection of seismic data and also deciding magnitude of the earthquake.

Environmental Issues and Future Scenario

After the 1972 Stockholm conference on Human Environment and Development, there has been a phenomenal increase in the worldwide interest in the problems related to the impact of developmental activities on environment. Consequently, the United Nations Environment Programme was created in 1972 for ultimately establishing a Global Environment Monitoring System (GEMS). This was followed by formulating various programmes to deal with climate, long-range transport of pollutants, terrestrial renewable natural resources, oceans and human health. The focus of development has shifted from exploitation to environmental friendly and sustainable developmental planning in recent years. To take stock of the situation a 12-day UN conference on Environment and Development (UNCED) known as Earth Summit was held at Rio de Janerio during June 3-14, 1992. It was a landmark development in the history of mankind, which resulted in the enunciation of the fact that earth is an integral and interdependent system, which require sustainable developmental planning. In the new millennium many policy makers, scientists, and common-man alike have started realizing that without the protection of our environment, there cannot be any worthwhile development. Through Agenda 21, it seeks to set the planet on a course to ecologically safe and sound development and save the 'Spaceship' Earth's fragile environment from further degradation.

It has been brought out vividly by the work done in India and elsewhere that satellite data is a very useful tool in documenting the state of the planet and provide exhaustive information on earth's resources and their dynamics. Another area where remotely sensed data is making significant contribution is related to global environmental change. During the present decade many remote sensing satellites have been/ are being launched, which will have improved spatial, spectral, radiometric resolutions. Thus, a vast amount of information flow will take place from the data from remote sensing satellites. This in conjunction with the information available from conventional sources can be integrated into a Geographic Information System (GIS), which will help in modelling various developmental scenarios for arriving at optimal solutions and ensuring sustainable development.

References

Anon. (2002), The State of Forest Report, Forest Survey of India, Govt. of India, Dehra Dun.

Garg, J.K., Singh, T.S. and Murthy, T.V.R. (1998). Wetlands of India. Project Report: RSAM/SAC/RESA/PR/01/98, June 1998, 240 p. Space Applications Centre, Ahmedabad,

Garg *et al,* (1990). Impact of Mining Activities and Super Thermal Power stations on Environment, Project Report: RSAM/SAC/RSAG/ENVN/PR/08/90, November 1990. Space Applications Centre, Ahmedabad, 100 p.

Jensen, J.R. (1986), Introductory Digital Image Processing: A Remote Sensing Prospective, pp. 234-252, Prentice Hall, Englewood Cliffs, New Jersey.

❑❑❑

GEOINFORMATICS APPLICATIONS IN AGRICULTURE, 2007
Eds. : A.K. Singh and U.K. Chopra; pp. 39-78
New India Publishing Agency, New Delhi (India)
E-mail : newindiapublishingagency@gmail.com
Web: www.bookfactoryindia.com

2

Geoinformatics in Precision Farming: An Overview

**Rajender Prasad, Anshu Dixit,
P K Malhotra and V K Gupta**

Introduction

Precision Farming (PF) revolves around the idea that any agricultural land contains wide spatial variations in soil types, nutrient availability, and other important factors. If these variations are not taken into account, then there may be a loss in productivity. Precision farming is one of the most scientific and modern approaches to sustainable agriculture that has gained momentum in 21st century. Consider the situation: 'A farmer goes to his/her field with his/her Global Positioning System (GPS) guided tractor. The GPS senses the exact location of the tractor within the field. It sends the signal to a computer on the tractor that has a Geographical Information System (GIS), storing the soil nutrient requirements map in it. GIS, in consultation with a Decision Support System, decides what is the exact requirement of the fertilizer for a given location and then commands a variable rate fertilizer applicator, attached with the tractor, to apply the exact dose at that precise location. All this is done within a second, before the tractor moves further.' This may sound like an excerpt from a scientific fiction, but this is what precision farming means to the large growers of the United States or European countries and day is not far off when it will be feasible for large Indian farmers as well.

PF refers to a developing agricultural management system that promotes variable management practices within a field according to site or soil conditions. Within a short span of time, this concept has been described with various names *viz.* farming by soil; farming soil, not fields; farming by the foot; spatially prescriptive farming; computer aided farming; farming by computer; farming by satellite; high-tech sustainable agriculture; soil-specific crop management; site-specific farming; and site-specific crop management.

PF employs a system engineering approach to crop production where inputs are given on an "as needed basis," and such a system has been made possible by recent innovation in information and technology such as microcomputers, GIS, GPS and automatic control of farm machinery. It is the holistic approach to micro management of spatial and temporal variability in agricultural land based on integrated soil, plant, information, and engineering management technologies as well as economics (Robert *et al.*, 1994).

As with any fledgling technology, PF has various definitions. Some define it as "...an information technology based agricultural management system to identify, analyze, and manage site-soil spatial and temporal variability within fields for optimum profitability, sustainability, and protection of the environment." whereas the others define it as "...a management strategy that uses information technologies to bring data from multiple sources to bear on decisions associated with crop production." A more precise and formal definition of PF is given in the sequel. According to Earl *et al.* (1997), "Precision Farming is the title given to a method of crop management by which areas of land/crop within a field may be managed with different levels of inputs depending upon the yield potential of the crop in that particular piece of land. It may help in reducing the cost of crop production in that area and the risk of environmental pollution from agrochemicals, applied at levels greater than those required by the crop.

Fundamentally, precision farming acknowledges those conditions for agricultural production as determined by soil resources, weather, and prior management that vary across space and over time. Given this inherent variability, management decisions should be specific to time and place rather than rigidly scheduled and uniform. Precision farming provides tools for tailoring production inputs to specific plots within a field, thus potentially reducing input costs, increasing yields, and reducing environmental impacts by better matching inputs applied to crop needs. Precision farming is a comprehensive system and involves almost all the disciplines of agricultural sciences, information technology, statistics and social sciences. It may not be possible to include details of the aspects related to each of the disciplines. The present chapter, however, attempts to present an overview of applications of geoinformatics in precision farming. We begin with the methodologies used in precision farming and the description of precision farming cycle in Section 2.

Precision Farming Methodology and Cycle

In this section, we briefly discuss the methodology for precision farming and precision farming cycle.

Methodology

Precision farming (PF) is comprised of numerous component technologies that include yield monitors, geo-reference grid soil sampling, geo-referenced variable rate technology for inputs application, GPS and detailed field maps created from GIS to name just a few. Following three technologies are the driving force for precision farming:

(i) global positioning systems (GPS) for positioning,

(ii) geographic information systems (GIS) for analyses, and

(iii) intelligent devices and implements (IDI)- such as yield monitors and variable-rate fertilizer rigs- for "on-the-fly" data collection and variable-rate control of field inputs.

(i) *Global Positioning System:* The global positioning system ("GPS") is a network of satellites developed for and managed by the United States Defence Department. The GPS constellation of 24 satellites orbiting the earth, transmits precise satellite time and location information to ground receivers. The ground receiving units are able to receive this location information from several satellites at a time for use in calculating a triangulation fix thus determining the exact location of the receiver.

(ii) *Geographical Information System:* A geographical information system ("GIS") consists of a computer software database system used to input, store, retrieve, analyze, and display, in map like form, spatially referenced geographical information.

(iii) *Intelligent Devices and Implements:* These include yield monitors and variable rate technology. **Yield monitors** are crop yield measuring devices installed on harvesting equipment. The yield data from the monitor is recorded and stored at regular intervals along with positional data received from the GPS unit. GIS software takes the yield and positional data and produces yield maps. **Variable Rate Technology (VRT)** is the method of controlling variability within the field. VRT consists of farm field equipment with the ability to precisely control the rate of application of crop inputs (fertilizer, insecticides, pesticides, herbicides, seed rate, plant population, and irrigation) and tillage operations in the field based on the individual characteristics of that location. Typical VRT system components include a computer controller, GPS receiver, and GIS map database. The computer controller adjusts the equipment application rate of the crop input applied. The computer controller is integrated with the GIS database. A GPS receiver is linked to the computer. The computer controller uses the location coordinates from the GPS unit to find the equipment location on the map provided by the GIS unit. The computer controller reads the instructions from the GIS system and varies the rate of the crop input being applied as the equipment crosses the field. The computer controller records the actual rates applied at each location in the field and stores the information in the GIS system, thus maintaining precise field maps of inputs applied.

Along with these three basic technologies, Remote Sensing (RS) technique holds great promise for precision farming because of its potential for monitoring spatial variability over time at high resolution (Moran *et al.*, 1997). Various workers (Hanson *et al.*, 1995) have shown the advantages of using RS technology to obtain spatially and temporally variable information for PF. RS imagery for PF can be obtained either through satellite-based sensors or video digital cameras on board small aircraft. Moran *et al.* (1997) summarized the applications of remote sensing for PF. According to them, RS can be used as a source of different types of information for PF. Nielsen *et al.* (1999) have described the opportunities for examining farm soil variability and suggested several alternatives for making soil maps for PF including use of spectral and co-spectral analysis. Department of Space, Government of India has already taken an initiative to generate soil resource maps at 1:50,000 scale for entire country using the IRS satellite and Linear Imaging Self Scanning (LISS) sensors. These soil resource maps will useful for delineating homogeneous zones with respect to soil fertility and crop yield and for estimating potential yield, yield gap analysis, monitoring seasonally -variable soil and crop conditions using optical and microwave optical data. Using RS data for mapping, however, has many inherent limitations *viz.* requirements for instruments calibration, atmospheric correction, normalization of off-nadir effects on optical data, cloud screening for data especially during monsoon period, processing images from airborne video and digital camera (Moran *et al.*, 1997). Keeping in view the agricultural scenario in developing countries, a marketable RS technology for PF is the delivery of information with the following characteristics *viz.* low turn around time (24-48 hrs), low data cost (approximately Rs.100/acre/season), high spatial resolution (at least 2m multi-spectral), high spectral resolution (<25nm), high temporal resolution (at least 5-6 data per season) and delivery of analytical products in simpler format.

RS, GPS and GIS technologies used in PF cover three aspects of production *viz.*data collection or information input, analysis or processing of the precision information, and recommendations or application of the information.

Data collection occurs both before and during crop production, and is enhanced by collecting precise location coordinates using GPS. Data collection technologies operating in advance of crop production include grid soil sampling, yield monitoring, remote sensing, and crop scouting. Other data collection takes place during production through "local" sensing instruments mounted directly on farm machinery. For example, soil probes mounted on the front of fertilizer spreaders can continuously monitor electrical conductivity, soil moisture and other variables to predict soil nutrient concentrations and to instantaneously adjust fertilizer application at the rear of the spreader. Optical scanners detect soil organic matter, or "recognize" weeds, to instantaneously alter the amount or application of herbicides applied.

Precise data are useless unless they can be analyzed or processed to enable management adjustments. GIS technology is used to integrate spatial data coming from various sources in a computer. This is primarily an intermediate step, because data collected at different times on the basis of different sampling regimes and different scales must be combined for use with subsequent decision technologies, such as process models, artificial intelligence systems, and expert systems. What makes PF different is the capability to capture data on the production practices actually applied in field and on the results achieved.

Precision Farming Cycle

Precision farming should not be thought of as only yield mapping and variable rate fertilizer application and evaluated on only one or the other. PF technologies will affect the entire production function (and by extension, the management function) of the farm. A brief overview of the precision farming cycle is presented in Figure 1.

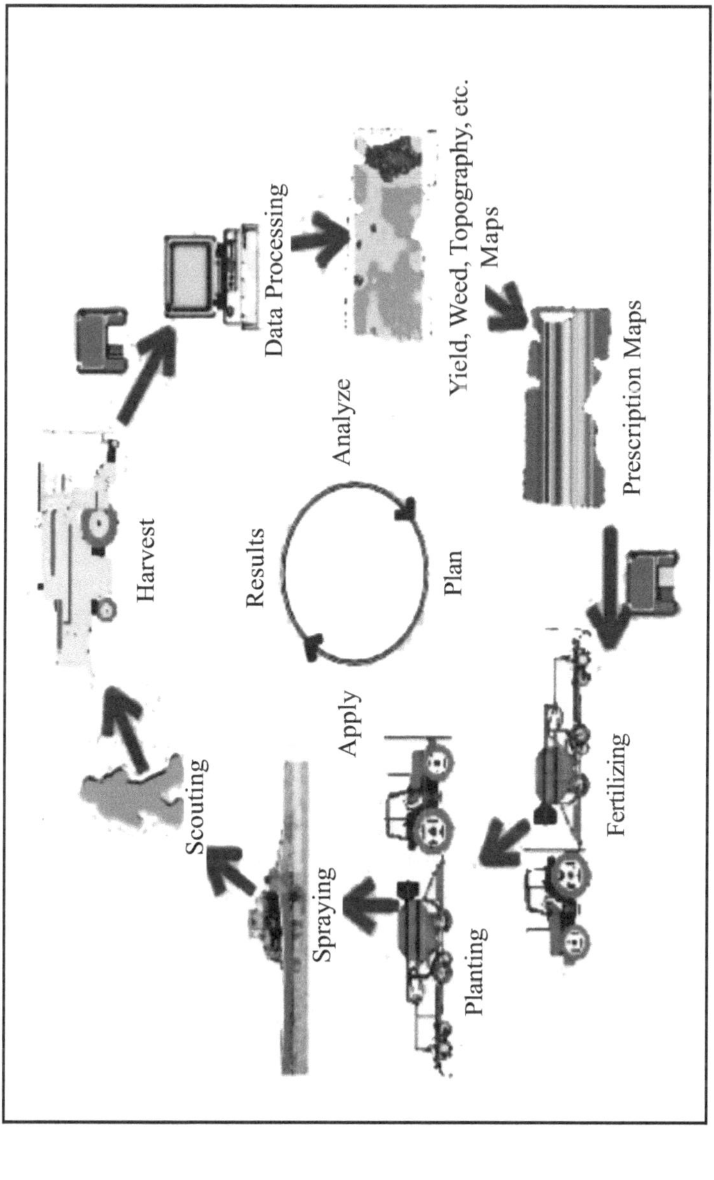

Figure 1. Precision Farming Cycle

[*Source:* http://www1.agric.gov.ab.ca/$department/deptdocs.nsf/all/sag1951?opendocument]

The various components used in the PF cycle are: a) Yield monitoring; b) Yield mapping; c) Variable rate fertilization; d) Weed mapping; e) Variable spraying; f) Topography and boundaries; g) Salinity mapping; h) Guidance system and i) Records and analyses.

The various production aspects & agricultural technologies for precision farming can be summarized as given in Table 1.

Table 1. The Suite of Precision Agriculture Technologies

Production aspect	**Technology**	
Data collection/input	Global positioning system (GPS) *In advance of production:* Grid soil sampling, yield monitoring, remote sensing, crop scouting	*During production:* Local sensing of : nutrients, pH, weeds
Analysis/processing	Geographic information systems (GIS), process models, artificial intelligence systems, expert systems, human decision-makers	
Recommendation/ application	*Variable-rate application:* Fertilizer, micronutrients, lime, herbicides, insecticides, seeds, seed variety, drip irrigation	*Selective harvest:* Harvest timing

Geoinformatics and Precision Farming

Geographical, earth sciences and agricultural sciences rely increasingly on digital spatial data acquired from remotely sensed images, analyzed by GIS and visualized on the computer screen or on paper. The technologies supporting the processes of acquiring, analyzing and visualizing spatial data form the core of geo-informatics (an interdisciplinary field requiring synergistic modelling and analysis for dealing with geospatial data). Geoinformatics offers three technical tools to administer and to monitor geographical information *viz.* GIS, GPS and Remote Sensing. It integrates various disciplines (Surveying, Remote Sensing, Cartography, Geography, Informatics, Computer Science and Statistics) dealing with geographical data (data linked to some locality on the surface of the earth). The key element that differentiates geoinformatics from other areas of information technology is the requirement that all data are geocoded, *i.e.*, has an address in 3-D space. The location of an entity provides us with a way to inter-relate a wide range of data types and discover possible relationships. Dixit *et al.* (2004) presented an overview of applications of geoinformatics in precision farming. In the present article, we elaborate on several issues given in Dixit *et al.* (2004).

Geoinformatics focuses on the theory and practice of the design and development of GIS and analytical methods that are typically required in a variety of applications including PF.

PF techniques may involve geostatistical maps and models, spatial models for plant diseases and arthropod population dynamics, incorporation of remotely sensed data, data collected on different spatial scales and with different measurement precision, and may use specialized sampling schemes taking advantage of or accounting for spatial correlations. PF data are collected on such small spatial scales that enormous data sets are generated. How are these data to be analyzed to extract as much information as possible? How should remotely sensed data, such as satellite imagery, be combined with crop yields measured at 3 meters intervals and with soil samples taken every 25 meters, each on different grids? Given that yields, disease incidence, and arthropod abundance are spatially correlated, how should one sample to determine critical levels and treatment responses? Many statistical questions need to be addressed in order to provide important information useful in performing agricultural management on precise scales. It is clear that raw data sets are too large to understand or interpret. Statistics provide a means of summarizing data and can be readily interpreted for making management decisions and can define relationships among variables.

To address these issues, a number of statistical tools *viz.* measures of central tendency like mean, median, mode; measures of dispersion, correlation and regression or the much talked about branch of applied statistics called geostatistics can be useful. *Measures of Central Tendency* are useful when a factor, such as crop yield, is measured at different locations within a field and values may vary greatly from location to location.

This variation is to be studied through geoinformatics. There may be a difference in the magnitude of measurements of a variable. Values can change randomly because of error in the sensor or they can change because of changes in the underlying factor *Temporal* or *Spatial*.

The set of measurements taken from different points in space is called a population and a measure of central tendency is the *central* or *usual* value of the population. This is important because *the current practices in PF* treat the field or area in the field based on the average of the measurements within the field or area. *Measure of dispersion d*escribes the distribution of the set of measurements. The most commonly used measures of dispersion are r*ange, standard deviation and coefficient of variation. Correlation and Regression* helps in studying the inter-relationships among the variables. In PF, one has to deal with geo-referenced spatial data, therefore, geostatistics is a potential tool for data handling, analysis, drawing valid inferences and taking right decisions in the areas where the data has the spatial aspect attached to it, including PF. Geostatistics in its original usage is referred to as Statistics of "earth" such as geography and geology. One of the essential features of geostatistics is that the phenomenon being studied takes values (not necessarily measured) everywhere in the study area. Geostatistics can be regarded as a collection of numerical techniques that deal with the characterization of spatial attributes, employing primarily random models in a manner similar to the way in which time series analysis characteristics temporal data. Both time series analysis and geostatistics are conceptually and historically related and primarily address the situation in which inferences must be drawn from auto-correlated data that are insufficient for obtaining precise results. Most geostatistical techniques rely on random variables to model the uncertainty associated with such assessments. It deals with the characterization of spatial attributes also known as regionalized variables (*the average over a certain volume or a surface rather than at a point. The basic volume on which a regionalized variable is defined is called its support)* for which deterministic models are not possible because of the complexity of the natural processes and often, the impossibility is compounded by high measurement cost.

Geostatistics procedures help us to recognize and describe spatial relationships that might exist in an agricultural field. To recognize spatial trends, the sample position is just as important as the measured value. The main techniques used in geostatistics analysis are structural analysis and interpolation.

Structural Analysis

It is the process of quantifying and modeling the spatial variability (a source of spatial uncertainty). It is nothing but the investigation of the spatial data structures. One key tool of structural analysis is *variography* which uses the structure function called *Variogram (semi-variogram)*. Here, variography is the process of estimating the theoretical semi-variogram. It begins with exploratory data analysis then computing the empirical semi-variogram, binning, fitting a semi-variogram model and using diagnostics to assess the fitted model. It is the simplest way to relate uncertainty with distance from an observation. Variogram is a function of the distance and direction separating two locations used to quantify autocorrelation and can be defined as the variance of the difference of the distance between two variables at two locations. The semi-variogram and covariance functions measure the strength of statistical correlation as a function of distance. Semi-variogram is a mathematical model of the semivariance as a function of lag. It displays the statistical correlation of nearby points. As the distance increases, the likelihood of these points being related becomes smaller. When two locations are close to each other, then they are expected to be similar. Their covariance is inversely proportional to the distance between locations. The variogram generally increases with distance. A variogram/semi-variogram may be described by

- Nugget - variability at zero distance, represents sampling and analytical errors
- Range - the extent of spatial trends, distance beyond which sampling is random
- Sill - variability of spatially independent samples

The variogarm /semi-variogarm use range to determine maximum sampling distances; the sill indicates intra-field variability and the model can be used for interpolation of values in unsampled areas. Mathematically, semi-variogram is described in the sequel.

Mathematical Description of a Semi-variogram

Consider two numerical values $Z(x)$ and $Z(x + h)$ at two location points x and $x + h$ separated by vector h. The variability between two quantities is characterized by the variogram function $2\gamma(x,h)$ which is defined as the expectation of the random variable $[Z(x)-Z(x+h)]^2$, *i.e.*,

$$2\gamma(x,h) = E[(Z(x) - Z(x+h))^2]$$

Here E(.) denotes the Expectation. In all generality, the variogram $2\gamma(x, h)$ is a function of both the point x and the vector h. For an intrinsic random function of order 0, $2\gamma(x,h)$ depends only on the separation vector h and not on the location x and denoted by $2\gamma(h)$. It is possible to estimate the variogram $2\gamma(h)$ from the available data. By taking arithmetic mean of squared differences between two experimental measures $[Z(x_i), Z(x_i+h)]$ at any two points separated by vector h

$$2\gamma(h)=\frac{1}{N(h)}\sum_{i=1}^{N(h)}[Z(xi)-Z(xi+h)]^2$$

where $N(h)$ is number of experimental pairs.A semi-variogram is a graphical display of $\gamma(h)$ versus lag (h), *i.e.*, semivariance versus distance. Usually, $\gamma(h)$ will increase with h, indicating more deviation and less correlation between Z-values with increasing distance. It is often the case that after a certain distance, called the range, *(a)*, $\gamma(h)$ will level out at a value called the sill $(C_0 + C_1)$. The range is, therefore, the distance beyond which the deviation in Z-values does not depend on distance and hence Z-values are no longer correlated. A typical graphical display of a semi-variogram is given in Figure 2.

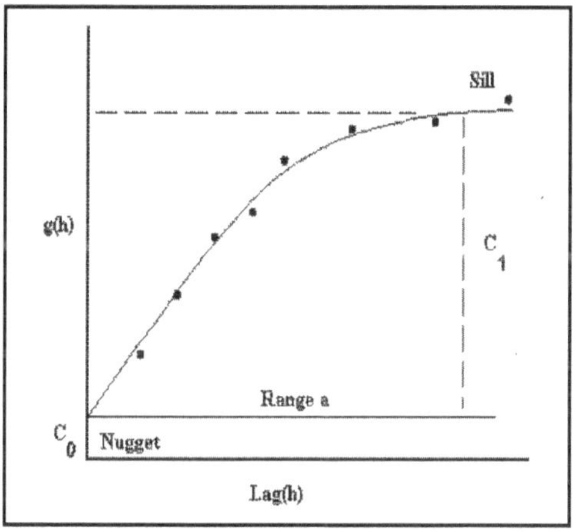

Figure 2. A semi-variogram

To help choose a *theoretical* semi-variogram model that fits a particular data, an *experimental* semi-variogram is constructed. To compute the experimental semi-variogram, data are divided into distance classes called lags. The number of lags and the size of the lags must be specified by the user. Estimation of g(*h*) is made for each lag.

To help determine the lag size, one has to be careful that the total distance over which g(*h*) is calculated is equal to the number of lags *times* the size of each lag (*h*) and that this distance should span at least half the data set.

The above description is that of a one-dimensional semi-variogram. There do exist two Dimensional Semi-Variogram and Covariance Function and Auto Correlation in two Dimensions.

Semi-variograms are of two types viz. isotropic and anisotropic depending on whether they depend on the direction or not.

Anisotropic semi-variogram: When the semi-variograms, obtained in different directions are different then that phenomenon is known as anisotropy and the semi-variograms are said to be anisotropic. In other words anisotropic semi-variograms are the function of *h* alone but also the direction (angle) in which they are obtained.

Isotropic semi-variogram: When the semi-variograms, obtained in different directions are same then that phenomenon is known as isotropy and the semi-variograms are said to be isotropic. In other words isotropic semi-variograms is function of h alone and independent of the direction in which they are obtained.

Knowledge of anisotropy is necessary when designing an appropriate semi-variogram model of the data prior to prediction at an unknown point.

Semi-variogram models

There are several models that can be fitted to the experimental semi-variogram. These models can be categorized by the presence or absence of a sill and by the behavior at the origin, which is either linear or parabolic. In addition, there may be a discontinuity at the origin, the so-called nugget effect. This means that the fitted model does not pass through the origin, but intersects the y-axis at a positive value of g(*h*), labeled C_0. The choice of theoretical semi-variogram is determined largely by experience and experimentation. Certain admissible variogram models are: linear, spherical, exponential, hyperbola, logarithmic, Gaussian, models without a sill and Hole-effect models. For detailed description on variography, one may refer to Cressie (1991) and Bhatia and Parsad (2003).

Interpolation

Interpolation can be defined as predicting values at locations where data has not been observed, using data from locations where data has been collected. Usually, interpolation is for predictions within area where data has been collected, rather than extending predictions to areas outside of the data collection area.

Spatial interpolation involves finding a function *f(x)* which best represents the entire surface of Z-values (called data *values*) associated with irregularly located *(x, y)* points (called data *sites*). In addition, this function predicts Z-values for other regularly spaced locations. Such a function is referred to as an interpolant. There are two types of interpolants, *exact* and *approximate* (data smoothing).

Interpolations techniques can be divided into two groups *viz.* deterministic and geostatistical. The deterministic interpolation technique is used for creating surfaces from measured points based either on the extent of similarity (e.g., inverse distance weighted [IDW]) or the degree of smoothing (e.g., radial basis functions and global and local polynomials). Deterministic interpolation techniques can furtherbe divided into two groups: global and local. Global techniques calculate predictions using the entire data set. Local techniques calculate predictions from the measured points within specified neighbourhoods, which are smaller spatial areas within the large study area.

An interpolator can either force the resulting surface to pass through the data values or not. An interpolation technique that predicts a value identical to the measured value at a sampled location in known as an exact interpolator. An inexact interpolator predicts a value at a sampled location that is different from the measured value. The ineact interpolator can be used to avoid sharp peaks or troughs in the output surface. IDW and radial basis functions are exact interpolators, while global and local polynomials are inexact or approximate interpolators

The *geostatistical interpolation technique* is used for more advanced prediction surface modeling that also include errors or uncertainty of predictions. The main geostatistical interpolation techniques are kriging and cokriging.

Kriging and co-kriging: The method of estimation embodied in regionalized variable theory is known in earth sciences as kriging after D.G. Krige who devised it empirically for use in the South African gold fields (Krige, 1966). It is essentially a means of weighted local averaging in which the weights are chosen so as to give unbiased estimates while at the same time minimizing the estimation variance. Kriging is in this sense optimal and generally called as a local estimation technique which provides the best linear unbiased estimator (BLUE) of the unknown characteristics studied. Kriging is the statistical interpolation technique that uses data from a single data type (single attribute) to predict (interpolate) values of same data type at unsampled locations. It also provides standard errors of predictions. Kriging is divided into two distinct tasks *viz.* quantifying the spatial structure of the data and producing a prediction

To make the prediction, kriging uses the fitted model from variography, the spatial data configuration, and the values of the measured sample points around the prediction location.

It is a moderately quick interpolator that can be exact or smoothed depending on the measurement error model. It is very flexible and allows the user to investigate graphs of spatial autocorrelation. Kriging uses statistical methods that allow a variety of indicators, and probability. The flexibility of kriging can require a lot of decision making. Kriging assumes the data comes from the stationary stochastic process. A stochastic process is a collection of random variables that are ordered in space and or time such as elevation measurements.

Co-kriging is the multivariate equivalent to kriging. For multiple datasets, it is a very flexible interpolation method, allowing the user to investigate graphs of cross-correlation and autocorrelation. When we want to use kriging with multiple datasets, then cross-variance models need to be developed. Cross variance is the statistical tendency of different types of variable to vary in ways that are related to each other. Positive cross variance occurs when both variables tend to be above their respective means together, and negative cross variance occurs if one variable tends to be above its mean when the other variable is below its mean. Cross variance modeling is used to define the local characteristics of spatial correlation between two datasets and used to look for spatial shifts in cross-correlation between two data sets. Co-kriging can use either semi-variogarm or covariance. It can use transformations and remove trends, and it can allow for measurement errors in the same situations as for the various kriging methods.

Thus in brief Kriging is a widely used method of spatial interpolation. Kriging is a minimum mean-squared error technique of spatial prediction that usually depends on the second-order properties of the process and is based variogram. Estimation and fitting of the variogram, as well as variogram model selection, are crucial stages of spatial prediction, because the variogram determines the kriging weights of the predictor. These three steps must be carried out carefully, otherwise kriging may produce non informative maps.

Regardless of the interpolation method there will always be some error in the analysis. This error can be influenced by many factors such as inaccurate base data, variation in physical properties of data samples, human error, and temporal, and temporal changes between measurements. All of these factors play a role in the severity of the measurement errors. *Error modeling* is used to minimize the effect of these measurement errors.

In order to describe a spatial trend within a field, multiple samples must be taken within the range of that trend. Samples from the field should be taken uniformly and should be representative of the variability in the field. Statistical designs for factorial experiments *viz.* uniform designs can be quite useful in selection of sites for taking samples. The uniform designs and their use with respect to PF are described in the sequel.

Uniform Designs and their use for sampling locations in PF

Uniform designs proposed by Fang (1980) and Wang and Fang (1981) are efficient and robust fractional factorial plans. Uniform design is one of space filling designs which *scatter points uniformly over* the experimental domain in uniform fashion. A uniform design is a design in which the design points distribute uniformly over the entire design space. The uniform designs can be defined as follows: consider that s factors are to be studied in an experiment via n experimental units. Without loss of generality, assume that the experimental domain is the unit cube $C^5 = [0,1]^s$; then a design consists of *n* points in C^5. We want these points to be uniformly scattered over C^5. A measure of uniformity called discrepancy is adopted, and the goal is to choose *n* design points with the smallest discrepancy. We denote an *n*-run design over C^5 by $X = (x_1, x_2, ..., x_n)'$ and let $P_{n,s}$ be the set of all such designs. Let

$$Fn(x) = \frac{1}{n}\sum_{i=1}^{n} I\{Xi < x\},$$

where I {.} is the indicator function and all inequalities are understood to be component-wise. The discrepancy of x is definedDp$[\int_{C^s} |F_n(X) - F(X)|^p \, dX]^{1/p}$, *where* F (x) is the distribution function of the uniform distribution over C^S. The design with smallest discrepancy. is called a Unfirom Design (UD). The above definition was given by Fang *et al.* (1999).

The concept of uniform designs was motivated by three big projects in system engineering in 1978. Uniform designs have been widely used in various fields, such as chemistry and chemical engineering, pharmaceuticals, quality engineering, system engineering, survey design, computer sciences and natural sciences.

Since these designs are space filling designs and scatter the points in uniform manner, therefore, if we relate the factors of variability with the factors of the factorial experiment and different levels of variability as the levels of the factors of the factorial experiment, then the design points of uniform design may be regarded as a representative sample of the field variability. Now one can select the samples at the locations identified through these points for measuring different parameters such as soil available nutrients, soil moisture, insect and pest incidence, etc. Once the observations on these parameters are taken and analyzed through variography, they can be used for spatial interpolation on the points not sampled in the field.

Another computer intensive technique that can be quite useful for precision farming is spatial data mining and is described in the sequel.

Spatial data mining for Site-specific Agriculture or Precision Farming

Large investments in technology and data collection are being made in the area of PF/variable rate application practices. Relatively little analysis on the utility of data is currently being performed. A richer set of analytical tools are needed to examine interactions between spatial and temporal characteristics of exogenous effects (weather), inputs (fertilizer, seed rate, pesticides, herbicides) and in-situ resources (soil characteristics, physiographic properties). Spatial data analysis tools help in evaluating agricultural practices to reduce in-field variability in an effort to maintain or improve crop yields. Spatial analytical tools are based on the facts that attributes of the neighbors of some object of interest may have an influence on the object and, therefore, have to be considered as well. The explicit location and extension of spatial objects define implicit relations of spatial neighborhood (such as topological, distance and direction relations) that are used by spatial data mining algorithms. Therefore, new techniques are required for effective and efficient data mining.

In spatial data mining focus is on assessing the feasibility of predicting the spatial yield map based on soil sample data (soil chemistry, physical properties), weather information (degree days, precipitation), planting date, cropping and management history, and other available information. An expected by-product of the analysis is an initial assessment of the potential benefits of variable rate application.

Subsequent work may incorporate web-based data serving and analysis tools under development to prototype an agricultural spatial analysis sub-system. This combination will greatly extend most spatial information systems from data repositories to an active analytical and modeling tool.

Technical Issues: Where to Start?

Now that the basic elements are in place four basic processing steps in PF tie it all together:

- Continuous data logging
- Point sampling
- Mapped data analysis
- Spatial modeling

PF does not "happen" as soon as a GPS unit or yield monitor is purchased. It occurs over time as a farmer adopts a new level of management intensity on the farm. Implicit in this is an increased level of knowledge of the PF technologies such as GPS. What is perhaps more important for the success of PF, at least initially, is the increased knowledge that a farmers' needs of his/her natural resources in the field. This includes a better understanding of soil types, hydrology, microclimates and aerial photography. A farmer should identify the variance of factors within the fields that effect crop yield before a yield map is acquired. A yield map should serve as verification data to quantify the consequences of the variation that exist in a field. Management strategies and prescription map development rely on sources other than yield maps.

Data logging continuously monitors measurements, such as crop yield, as a tractor moves through a field. Point sampling, on the other hand, uses a set of dispersed samples to characterize field conditions, such as phosphorous, potassium, and nitrogen levels. The nature of the data derived by the two approaches are radically different- a "direct census" of yield consisting of thousands of on-the fly samples versus a "statistical estimate" of the geographic distribution of soil nutrients based on a handful of soil samples as mentioned in Section 3.

In data logging, issues of accurate measurement, such as GPS positioning and material flow adjustments, are of major concern. Most systems query the GPS and yield monitor every second, which at 4 mph translates into about 6 feet. With differential positioning the coordinates are accurate to about a

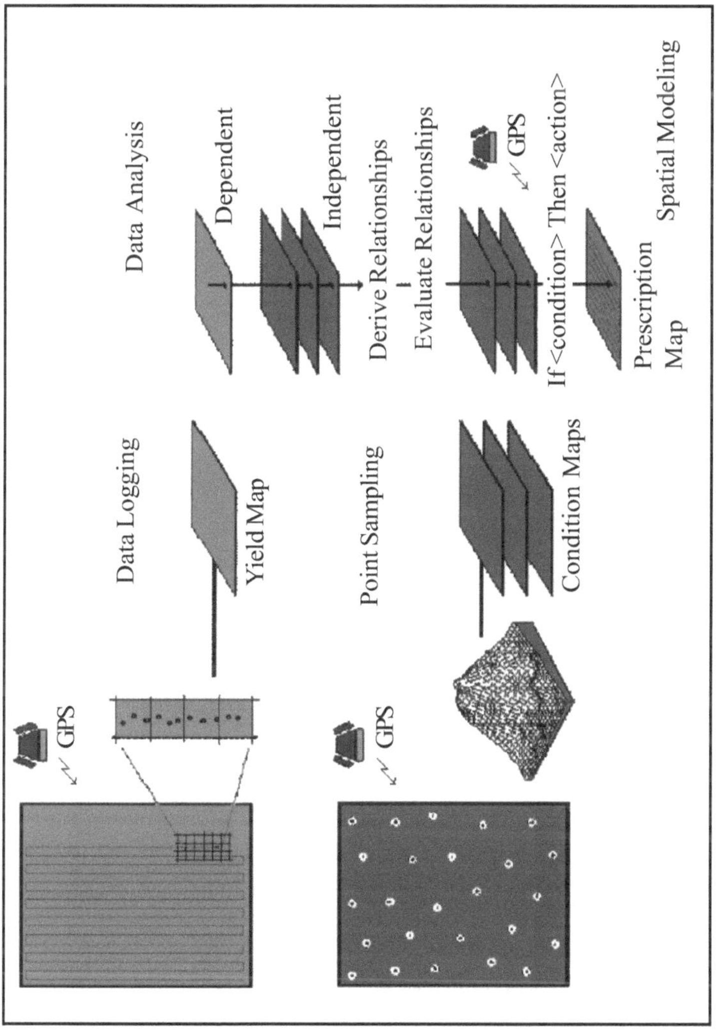

Figure 3. Flowchart of the PF process.
Source: http://www.innovativegis.com/basis/pfprimer/Topic3/Topic3.htm

meter. However, the paired yield measurement is for a location well behind the harvester, as it takes several seconds for material to pass from the point of harvest to the yield monitor. To complicate matters, the mass flow and speed of the harvester are constantly changing when different terrain and crop conditions are encountered. The precise placement of GPS/ Yield records are not reflected as much in the accuracy of the GPS receiver as in "smart" yield mapping software.

In *point sampling,* issues of surface modelling (estimating between sample points) are of concern, such as sampling frequency/pattern and interpolation technique. The cost of soil lab analysis dictates "smart sampling" techniques based on terrain and previous data be used to balance spatial variability with budget. In addition, techniques for evaluating alternative interpolation techniques and selecting the "best" map using residual analysis are available in some of the soil mapping systems.

In both data logging and point sampling, the resolution of the analysis grid used to geographically summarize the data is of critical concern. Like a stockbroker's analysis of financial markets, the fluctuations of individual trades must be "smoothed" to produce useful trends. If the analysis grid is too coarse, information is lost in the aggregation over large grid spaces; if too small, spurious measurement and positioning errors dominate the information.

The technical issues surrounding *mapped data analysis* involve the validity of applying traditional statistical techniques to spatial data. For example, regression analysis of field plots has been used for years to derive crop production functions, such as corn yield (dependent variable) versus potassium levels (independent variable). In a GIS, you can use regression to derive a production function relating mapped variables, such as the links among a map of corn yield and maps of soil nutrients—like analyzing thousands of sample plots. However, technical concerns, such as variable independence and autocorrelation, have yet to be thoroughly investigated along with statistical measures assessing results of the analysis, such as a spatially responsive correlation coefficient, await discovery and acceptance by the statistical community.

In theory, *spatial modeling* moves the derived relationships in space or time to determine the "optimal" actions, such as the blend of phosphorous, potassium and nitrogen to be applied at each location in the field. In current practice, these translations are based on existing science and experience without a direct link to data analysis of on-farm data. For example, a prescription map for fertilization is constructed by noting the existing nutrient levels (condition) then assigning a blend of additional nutrients (action) tailored to each location forming a if-(condition)-then-(action) set of rules. The issues surrounding spatial modeling are similar to data analysis and involve the validity of using traditional "goal seeking" techniques, such as linear programming or genetic modeling, to calculate maps of the optimal actions.

Summing up the above description, we can say that the *data logging* and *point sampling* steps provide the base maps of yield and field conditions, such as soil properties and nutrients. The *mapped data analysis* step uses the computer to establish the relationships between the output (yield map) and the inputs (soil properties and nutrient maps). This step is analogous to the agricultural researcher's derivation of the familiar crop production functions published in texts and reports. The main difference is that the analysis generates a production curve for a particular field, not for a research field 70 miles away on a different soil and using different varieties. The *spatial modeling step* uses thetailored "backyard" relationships to determine appropriate management actions and, it is , the "if <condition> then <action>" guidance a farmer needs. The difference from traditional research and guidance is that PF's recommendations respond to the unique conditions and variability within the given field and is reported in the form of a map. Admittedly, this process might seem a bit overwhelming at first- and more appropriate for a "techno-farmer"-but it could be heralding the next revolution in agriculture.

Crop Models in PF

Significant technical advances have been made in measuring and displaying variation in crop yields across a field, but it is not always clear how to determine the best management practice for each part of a field in order to achieve a particular goal. For example, does a farmer apply more fertilizer to the lower yielding areas to try and raise yields to the average, or are these low-yielding areas at their potential yield already, and would s/he, therefore, be advised to apply more to the high yielding areas on the basis that they are able to make better use of it? Similarly, are the optimal fertilizer application rates same for all parts of the field year after year, or do they vary in response the changing environmental conditions between years?

Answers to these questions are not clear-cut, but can be found using optimization techniques based on knowledge of the marginal yield response to a unit of the particular input for all the different regions of a field. The marginal yield response can be obtained from the curve describing yield response to the level of a particular input for each homogeneous region of the field. However, in most cases the response curves of each of these regions are unlikely to be known. While mini-experiments on each homogeneous region with a range of application levels of the input will provide this information, this approach is time-consuming, labour-intensive, and results are likely to be specific to that field only. Moreover, if responses change from year to year, then these experiments would have to be repeated every season.

Several ways are being explored in which crop models can be used to determine relationships between yields and levels of inputs, and to use this information to develop maps of the optimal distribution of the input to achieve specified goals such as yield maximization, input minimization, or minimization of environmental impact

An alternative, and perhaps complementary, approach is to use crop simulation models to predict the likely yield response to different levels of a particular input. Such models offer a cost-effective way in which agronomic knowledge accumulated from numerous previous experiments, usually with treatments of uniformly applied inputs on small and relatively homogeneous plots, can be extended to larger spatially-variable fields. Crop simulation models or crop simulators depict the symbiosis of statistics and computer science (IT) in PF. Crop simulators are computer programs that mimic the growth and development of crops. Data on weather, soil, and crop management are processed to predict crop yield, maturity date, and efficiency of fertilizers and other elements of crop production. The calculations in the crop models are based on the existing knowledge of the physics and physiology of plant responses to the environment. The farmers may use the various crop simulation models for the selection of cultivar/ soil type combination, planting date, row spacing, irrigation scheduling, harvest timing and yield prediction.

Application of Geoinformatics in PF: Some Case Studies

At all the levels, possible Geoinformatics applications in precision agriculture are almost endless. For example:

- To decide on the best potential sites for growing a certain cash crop, the *agricultural planner* might use geographic databases combining soils, topography and rainfall to determine the size and location of biologically suitable areas, and then overlay this with landownership and transport infrastructure, labour availability and distance to market centres.
- Further, one can study the changes in the characteristics of various attributes over time to determine the probable impacts of changing circumstances, such as the effects of a drought, the rise or fall of domestic or world prices, or the development of additional roads. Modeling, often involves proximity analysis and interpolation.

Geoinformatics has numerous applications in PF. To be specific some of them are described in the sequel.

- Site specific crop management:
  - Crop classification
  - Positioning guidance
  - Weed control
  - Crop protection and management
  - Forecasts for diseases
  - Soil characteristics studies
  - Fertilizer application
  - Irrigation application
- Soil study:
  - Assessment of soil erosion in a particular area
  - Soil suitability criterion search for different crops
  - Soil mapping
  - Soil fertility and nutrient management and site specific management
  - Soil test variability
  - Sustaining soil productivity with fertilizer management
- Weed and pest management:
  - Bio-economic weed management
  - Monitor the weed biology and ecology Mapping weeds
  - Studying herbicide adsorption and bio-availability in soils
  - Integrated pest management.
  - Variable rate herbicide application
- Post Harvest Control Operations
- Assessment and management of water logging areas
- Irrigation scheduling using RS
- Variable application of water and chemicals with sprinkler attached device

- Variable rate seeding
- Analyzing agricultural spatial data using GIS
- Land cover and land use change analysis specific to an area
- Searching best potential sites for growing a certain crop

The applications for finding the soil suitability criterion and site specific soil characteristics mapping and prediction are described in Sections 6.1 and 6.2 respectively.

Soil suitability criterion for growing a certain crop

Finding out the soil suitability criterion for a specific crop production in a particular geographical area is an important application of geoinformatics in PF. It is a multistage exercise that starts with the generation of soil suitability map. The suitability map is prepared considering the soil requirements for the cultivation of a specific crop. The important pedo-climatic requirements are studied and listed.

Pedo-climatic requirements of soil are different for different crops. The physico-chemical properties of soils such as depth, texture, available water capacity (AWC), pH, cation exchange capacity, etc. are the important soil characteristics in land evaluation that affect the yields under specific climatic and site conditions. The combined effects of these parameters are to be looked into, in a holistic approach rather than comparing individual characteristics with yield. In order to have realistic idea about yield contributing soil characteristics / qualities, data on level of crop management and the socio-economic data are required. Jack (1999) and Sys *et al.* (1991) have categorically stated that deriving soil suitability zones is a "Multi Criteria Decision Making" problem. The criteria / parameters controlling the soil suitability are soil depth, drainage, texture and available water capacity, each of which in turn contain different categories. These criteria and sub-criteria are of varying importance to the decision-maker. Consequently, information about the relative importance of the criteria is required. This is achieved by assigning a weight to each criterion. The derivation of weights is the central step to elicit the decision maker's preferences. After the weightages are derived, these evaluation criteria have to be integrated using multicriteria decision rules. The decision rules provide the basis for ordering the decision alternatives and for choosing the most preferred alternative.

The GIS based method for solving Multi Criteria based Decision Making problems involves the following steps:

1. Define the set of evaluation criteria (map layers) and the set of feasible alternatives.
2. Standardize each criterion map layer.
3. Define the criterion weights; that is a weight of "relative importance" is directly assigned to each criterion map.
4. Construct the weighted standardized map layers; that is multiplying the standardized map layers by the corresponding weights.
5. Generation of the overall score for each alternative using the add overlay operation on the weighted standardized map layers.
6. Rank the alternatives according to the overall performance score; alternatives with higher scores are most suitable or preferred.

Ahmed *et al.* (2003) have developed suitability index for potential agro-forestry areas in Yamunanagar district of Haryana, using the "Spatial Analytical Hierarchical Process (Spatial AHP)". In AHP (see e.g. Keeney and Raiffa, 1976; Yu, 1985; Zeleny, 1982) the decision maker is first asked to establish a hierarchy with at least two levels for evaluating candidate alternatives. Each level includes many competing criteria, each of which is assumed to be linearly independent from one to another. At the second step, the decision maker should subjectively give his/her perceived positive pair-wise comparison matrices (Hwang and Yoon, 1981), each of which should pass the consistent test (Leung and Cao 2000; Saaty,

1977). Thirdly, the decision maker is supposed to judge the relative importance (weight) among many competing criteria in each level and finally the hierarchical additive weighing method is usually used to find the priority of each alternative (Hwang and Yoon, 1981). Dixit *et al.* (2004) carried out an exercise to find the soil suitability criterion for coffee plantation (variety *Coffee Arabica)* in Paderu Mandal of Vishakhapatnam District of Andhra Pradesh. The study area for this exercise is a part of Paderu Mandal and consists of only 7500ha. Therefore, no appreciable differences in climatic conditions were expected. RS imagery from IRS-IB, LISS-II sensors A1, A2, B1 and B2 in the form of digital and geocoded false colour composite (FCC) covered by path 22 and row 55 of the February 26, 1993 was used for the study. Along with the remote sensing image, the Survey of India topographical map on 1:50,000 scales *viz.* 65J/12 was used to prepare the base map and to provide relevant ground truths. The ancillary data used is the information from published reports and maps. For the present study "ERDAS Imagine" version 8.4 was used to process the satellite data and apply GIS techniques. Information of physiography and analytical data were also collected and the maps were composed using ARC/INFO. An attempt was made through this study, first, to find out the suitability criteria and evaluating the soils

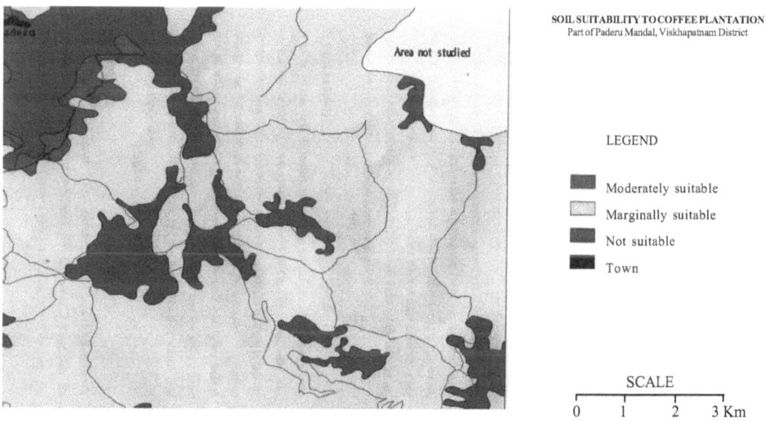

Figure 4. Soil Suitability Map for Coffee Plantation in Paderu district in Andhra Pradesh

for the development of coffee plantation and then to evolve an optimum land use plan based on different resource maps of the study area to develop an action plan. The soil mapping units were evaluated for suitability for coffee from different physio-chemical properties using "worst criteria principle". The resource maps of soil, land use and slope were merged and based on the information on physiography and ground truth, optimum land use plan was generated using GIS technique to obtain the optimum land use maps. The soil suitability map that was generated is given below.

A majority of the area was found "marginally suitable" to grow coffee (5860 ha). A small area of 187 ha was found "moderately suitable" while the remaining 1429 ha area was "not suitable" to raise coffee.

This approach takes into consideration various elements of natural resources, *viz.* all base layers, existing land use / land cover, soil map, hydrogeomorphology, terrain slope, drainage, watershed maps and local peoples' aspirations with respect to socio economic considerations to suggest a suitable action plan for sustainable development of land and water after scientific analysis of the spatial and non-spatial data. Therefore, it can be very useful for searching best potential sites for growing a certain crop.

Site specific soil characteristics and their prediction at a given point

Applications of agricultural inputs at uniform rates across the field without due regard to in-field variations in soil fertility and crop conditions does not yield desirable results in terms of crop yield. As mentioned earlier, the management of in-field variability in soil fertility and crop conditions for improving the crop production and minimizing the environmental impact is the crux of PF. Thus, the information on spatial variability in soil fertility status and crop conditions is a pre-requisite for adoption of PF. By catering to this variability one can improve the productivity or reduce the cost of production and diminish the chance of environmental degradation caused by excess use of inputs (Pierce and Nowak, 1999). Thus, mapping and analysis of within field variability is an essential input for precision crop management. Thus, PF involves acquiring the variations in crop or soil properties, mapping, and analyzing the variations, adopting suitable management techniques to maximize the yield. Farmers have been applying fertilizers based on recommendations emanating from research and field trials under specific agro-climatic conditions, which have been extrapolated to a regional level. Since soil nutrient characteristics vary not only between regions and between farms but also from field to field (Ladha *et al.*, 2000), and within a field, hence there is a need to take into account such variability while applying fertilizers to a particular crop. PF embodies the practice of applying crop inputs in each field or its parts according to its unique conditions.The information for variability map can be obtained from soil tests for nutrient availability, yield monitors for crop yield, soil samples for organic matter content, information in soil maps, or ground conductivity meters for soil moisture (Mulla, 1997). Cluster analysis can usefully be employed in classification of field plots into different categories based on data on several characteristics. Nigam *et al.* (2004) have used cluster analysis in the preparation of fertility contour maps using the uniformity trial data. The fertility maps and yield contour maps can also be generated using "Smart" yield mapping software, ARC GIS, etc. Singh and Elamaithi (2003) have generated the fertility (% organic carbon, Phosphorus, Potassium) and yield contour maps using ARC INFO under the project entitled *Development of Site Specific Management through PF Technologies in Rice-Wheat Cropping System* at Allahabad Agricultural Institute, Allahabad. For generating the contour map on organic carbon, one may consider different classes as <0.15, 0.15-0.30, 0.30-0.45, 0.45-0.60, 0.60-0.75. For phosphorus and potassium, the classes considered may be <0.25, 25-50, 50-75, 75-100 and 100-125 P_2O_5 kg/ha and <225, 225-250, 250-275, 275-300, 300-325 and 325-350 K_2O kg/ha respectively. The spatial statistics tools are also quite useful in studying the field variability. Bhatia *et al.* (1994) and Bhatia and Parsad (2003) have used the spatial statistics in studying the large field variability in hilly and salt affected soil regions.

Generally, the fields are manually sampled along a regular grid and the analyzed results of the samples are intrapolated using geostatistical techniques. These techniques are time consuming, labour intensive and in many cases destructive especially, for agricultural situation in India. With small size of landholdings and low income of farmers, the adoption of this methodology in its present form is not feasible. Various workers (Hanson *et al.*, 1995, Taylor *et al.*, 1997, Moran *et al*, 1997) have shown the advantages of using RS technology to obtain spatially and temporally variable information for PF. In an earlier work, Ray *et al*. (2001) have shown the usefulness of IRS merged data in mapping the variability.

For studying the soil characteristics specially the soil variability behaviour over a smaller grid, the geostatistical techniques like kriging can usefully be employed for predicting the soil characteristics at a given point. Such kind of problems also occurs in designed field experiments particularly in long-term fertilizer experiments. These experiments are generally carried out to study the long-term effects of given treatments and crops on soil fertility. The same design layout including randomization is followed over years. The data on soil available nitrogen, phosphorus, potassium, organic carbon, etc. is collected from each of the plots after harvest of the crop in each of the seasons. Soil analysis is generally performed in the laboratory and is quite time consuming. It has been observed that there is a data gap of more than one year in respect of soil available nutrients. Geoinformatics can be used for reducing this data gap by taking soil test values for some of the plots of the field layout and predict the soil status of the remaining plots by using the technique of variography and kriging. The plots selected for observing the soil status may be chosen in such a fashion that data from each of the plots is physically collected at least once in two to three years. For the feasibility study of the above, the data from a Long-Term Fertilizer Experiment (LTFE) conducted at Ludhiana on maize-wheat-cowpea sequence on soil available nitrogen under the aegis of All India Co-ordinated Research Project on Long Term Fertilizer Experiments has been taken. The experiment was

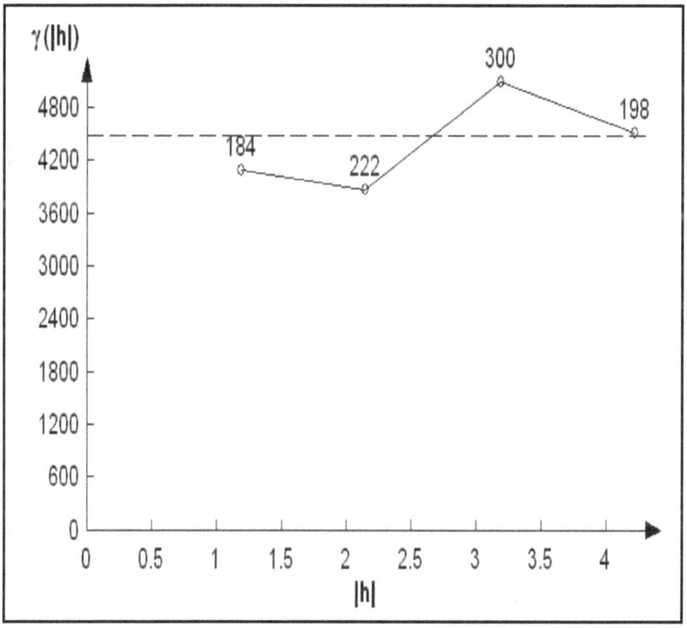

Figure 5: Variogram of nitrogen availability in soil

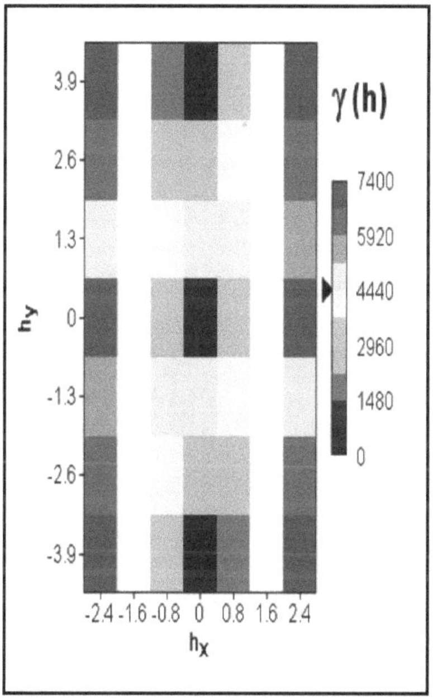

Figure 6. Grid generated for nitrogen availability in the soil

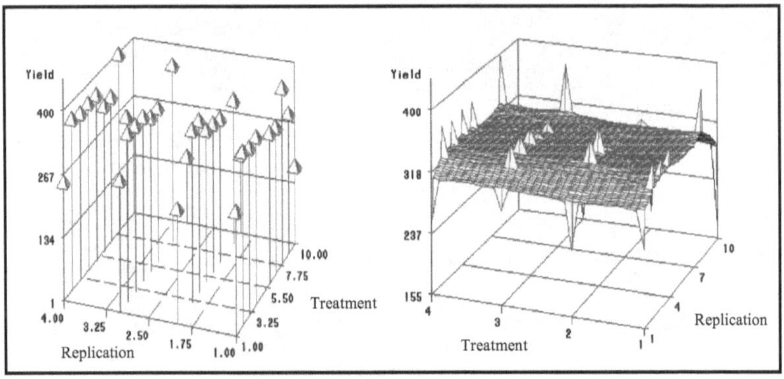

Figure 7 & 8. Surface plots generated for soil nitrogen

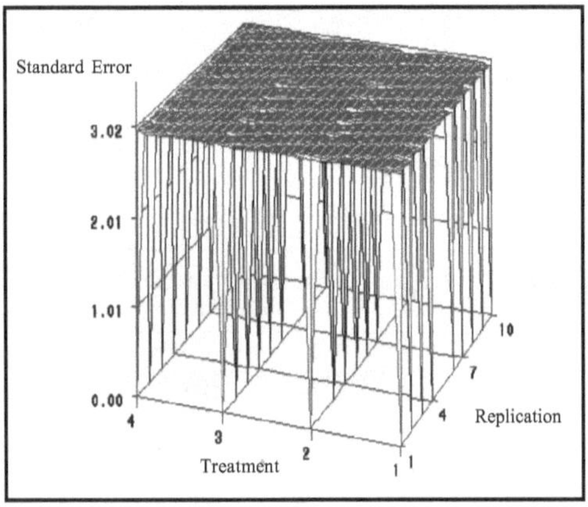

Figure. 9. Surface plot of standard errors of kriging estimates for soil nitrogen

conducted with 10 fertilizer treatments each replicated 4 times using a Randomized Complete Block design at Ludhiana. There were a total of 40 observations.10 observations were randomly dropped and then kriging was applied to predict the values at the locations from where the observations were dropped. The spatial analysis and interpolation are carried out using VARIOWIN and Statistical Analysis Systems (SAS). The variograms were prepared using VARIOWIN and the value of nugget, sill and range were calculated. These values of nugget, sill and range were used to create the surface plot and also to generate the grid; these operations were carried out in SAS. The variogram and krigged maps are given in the sequel.

The kriged variances were found to be less than 5%. Further studies are required to predict the soil behaviour at a given point with small prediction errors, *i.e.,* the actual differences between observed value and predicted value at a point.

Discussion

Information is perhaps the modern farmers' most valuable resource. Timely and accurate information is essential in all phases of production from planning through post harvest. Information available to the farmer includes crop characteristics, soil properties, fertility requirements, weed populations, insect populations, plant growth response, harvest data, and post harvest processing data. It is said "Precision farming can be referred to as the "original" form of agriculture". PF is changing the way in which agricultural research can be accomplished. The generation of massive amounts of data on farms will enable dynamic experimentation that could supercede the use of traditional controlled experimental plots. Information technologies can produce quantitative data that will complement qualitative whole-farm case studies. On-farm research will reflect actual farming practices. Further, the agricultural system may need to be evolved so that innovation and learning can exploit both traditional research plot experiments and information captured from actual field operations. Farmers engaged in PF are likely to be transformed from research clients into research partners. PF has altered our perception about what is technologically and professionally correct in plant production.

Geoinformatics is a defining opportunity for agriculture science research. The future research direction and opportunities in agriculture sciences will be significantly affected both by the availability and utilization of Information Technology. Now there is a need to adopt new tools, techniques and technology for proper use and conservation practices for resource use. Geoinformatics have evolved as a proven tool for resource planning. The high resolution satellite data and the

GPS use made it possible to plan at land parcel level. PF models and judicious use and regeneration of resources can be ensured at local level, at almost real time basis. Geoinformatics has proven as a powerful tool for studying the natural resources and help in locational features, extent of coverage, monitoring the resources and in generating models for probable scenarios, which assist in optimizing resource utilization. Experience has shown that data alone are insufficient to most users. "Users want solutions, not data" and "Users want information they can use immediately in a form they can understand" (Turner, 2002). Thus geoinformatics holds a great promise for resource planning in Indian agriculture in foreseeable future.

Resource planning is dependent on a spatial data management system that gives the user powerful decision making tools and is known as spatial decision support system (SDSS). The SDSS must include tools for seamless storage, retrieval, manipulation, and analysis of geographically referenced agricultural data. These "site-specific tools" are needed to systematically integrate the various components of PF technologies, from farm field to SDSS, back to farm field. Just having information about variability within the field doesn't solve any problem unless there is some kind of decision support system ("DSS") in order to make VRT recommendations. Russo and Dantinne (Russo *et al.*, 1997) have suggested the following steps for a SDSS:

1. Identify environmental and biological states and processes in the field that can be monitored and manipulated for the betterment of crop production.
2. Choose sensors and supporting equipment to record data on these states and processes.
3. Collect, store and communicate the field-recorded data.
4. Process and manipulate the data into useful information and knowledge.
5. Present the information and knowledge in a form that can be interpreted to make decisions.

Therefore, SDSS needs to be designed to help farmers solve complex spatial problems and to take decisions concerning to irrigation scheduling, fertilization, use of crop growth regulators and other chemicals.

To sum up, we can say that the scientific fact in PF strategy is the application of agricultural inputs based on crop demand soil attributes rather than applying at uniform rate across the field. This helps in optimization of growth condition, enhancement of fertilizer and biocide use efficiency. Further, the PF in many developing countries including India is in its infancy but there are numerous opportunities for adoption. According to Shanwad *et al.* (2004) progressive Indian farmers, with guidance from the public and private sectors, and agricultural associations, will adopt it in a limited scale as the technology shows potential for raising yields and economic returns on fields with significant variability, and for minimizing environmental degradation. In many developing countries, farmers will not adopt precision farming unless it brings in more or at least similar profit as compared to traditional practice. The support from governments and the private sector during the initial stages of adoption is, therefore vital. It must be remembered that not all elements of precision farming are relevant for each and every farm. For instance, introduction of variable rate applicators is not always necessary or the most appropriate level of spatial management in Indian farms. Likewise, not all farms are suitable to implement precision farming. Some growers are likely to adopt it partially, adopting certain elements but not others. The adoption of PF can be improved with effective coordination among the public and private sectors and growers.

References

Ahmed, T., Singh, R. and Rai, A. 2003. *Development of GIS based technique for identification of potential agro-forestry area.* Project Report, IASRI, New Delhi.

Bhatia, V.K. and Parsad R. 2003. *Studies on data processing techniques for statistical analysis of large field variability in hilly and salt.* NATP-CGP Project Report, IASRI, New Delhi.

Bhatia, V.K., Samra, J.S. and Narain, P. 1994. *Studies on spatial patterns and its role in analysis of agricultural field experiments.* Project Report, IASRI, New Delhi.

Bootlink, H.W.G., J.Verhagen, J.Bouma, P.K. Thornton, 1996. Application of simulation models and weather generators to optimize farm management strategies. In *Proceedings of the 3rd International Conference on Precision Agriculture.* June 23-26, 1996 Minneapolis, Minnesota 343-360.

Cressie, N. (1991). *Statistics for Spatial Data.* John Wiley, New York.

Dixit, Anshu, Kotur, S.C, and Ravi Sankar, T. 2004. Application of remote sensing and GIS in the management of agricultural resources: A case study in a part of Paderu mandal of Vishakhapatnam Distt. Andhra Pradesh. *Computers and Electronics in Agriculture* (Submitted).

Dixit, Anshu, Parsad, R. and Malhotra, P.K. 2004. Applications of Geoinformatics in Precision Farming: An Overview. In *Proceedings of the Symposium on Geoinformatics Applications for Sustainable Development,* IARI, New Delhi, Feb. 17-19, 2004, 115-125.

Earl, Wheeler, Blackmore, 1997. Precision Farming - The Management of Variability. *The Journal of the Institution of Agricultural Engineers, Landwards,* vol. no. 4 pp18-23, The journal of the Institution of Agricultural Engineers, July 16, 1997. Available: http://www.silsoe.cranfield.ac.uk/cpf/papers/Landwards/LANDW-P_FinalF.htm

Fang, K.T., Shiu, Wai-Chee and Pan, Jian-Xin. 1999. Uniform designs based on Latin squares. *Statistics Sinica,* 9, 905-912

GopalaPillai, S., Tian, L. 1999. In-field variability detection and spatial yield modeling for corn using digital aerial imaging. *Transactions of the ASAE.,* 42(6), 1911-1920.

Hanson, L.D., Robert, P.C. and Bauer, M. 1995. Mapping wild Oats infestation using digital imagery for site-specific management. In *Proceeding Site-Specific Management for Agricultural System.* March 1994, Minneapolis, MN, ASA-CSA-SSSA, Madison, WI, 495-503

Hwang, C. L. and Yoon, K. 1981. in *Economics and Mathematical Systems: Multiple Attribute Decision Making, Methods and Application, A State of Art Survey,* Lecture Notes, Springer-Verlag.

Jack, M. 1999. *GIS and Multicriteria Decision Analysis.* John Wiley & Sons, New York, 392 pp.

Keeney, R. L. and Raiffa, H. 1976. *Decisions with Multiple Objectives, Preferences and Value Trade offs*. Cambridge University Press.

Krige, D.G. 1966. Two-dimensional weighted moving average trend surfaces for ore valuation. *In proceedings of the symposium on Mathematics, Statistics and Computer Application in ore valuation, Johannesburg*. South African Institute of Mining and Metrallurge, Johannesburg.

Leung, L. C. and Cao, D. 2000. On consistency and ranking of alternatives in fuzzy AHP," *European Journal of Operational Research*, 124(1), 102-113.

Mulla, D.J. 1997. Geostatistics, remote sensing and PF. In *Precision Agriculture: Spatial and Temporal Variability of Environmental Quality*, Eds. J.V.Lake, G.R. Bock and J.A. Goode, John Wiley & Sons, New York, 100-119.

Moran, M.S., Inoue, Y. and Barnes, E.M. 1997. Opportunities and limitations for image-based remote sensing in precision crop management. *Remote Sensing Env.*, 61, 319-346

Nielsen, D.R.; Wendroth, O.; Pierce, F.J. 1999. Emerging concepts for solving the enigma of precision farm research. In: International Conference On Precision Agriculture, 4., Madison. Proceedings, 303-318.

Nigam, A.K., Parsad, R. and Gupta, V.K. 2004. *Design and Analysis of On-station and On-farm Agricultural Research Experiments: A Revisit*. Project Report, ICAR AP-Cess Fund. Joint Publication of IASDS, Lucknow and IASRI, New Delhi.

Pierce, F.J., Nowak, P. 1999. Aspects of precision agriculture. *Adv. in Agron.*, 67, 1-85.

Ray, S.S., Panigrahy, S. and Parihar, J.S. 2001. Precision farming in Indian Context. *GIS@development*, November 2001.

Robert, P.C., Rust, R. H. and Larson, W.E. 1994. Site-specific management for agricultural systems. *American Society of Agronomy*, Preface p. 13, Madison WI.

Russo, J., and Dantinne, J., 1997. *Seeing the Trees and the Forest. Precision AG Illustrated*, July-August 1997, 1(4), 31-32

Saaty, T.L. 1977. A scaling method for priorities in hierarchical structures. *Journal of Mathematical Psychology*, 15(3), 234-281.

Singh, S.S. and Elamaithi, S. 2003. *Development of site-specific management through precision farming technology in rice-wheat cropping system.* Research project of Department of Agronomy, Allahabad Agricultural Institute, Allahabad.

Sys, Ir C, Van Rast, E., Debaveye, Ir J. and Boernaert, F. 1993. Land Evaluation. Part III Crop requirements. Agricultural Publication No. &, General Administration for Development Cooperation Place du Champ de Mars 5 bte 57 1050, Brussels, Belgium.

Taylor, J.C., Thomas, G. and Wood, G.A. 1997. Diagnosing sources of within-field variation with remote sensing. In *Precision Agriculture' 97*, Vol. II: Technology, IT and management (Ed. J.V. Stafford). BIOS Scientific Publishers Ltd., Oxford, UK. 705-712.

Turner, A. Keith <mailto: kturner@mines.edu> 2002. Geoinformatics: Providing geoscience information to satisfy user needs. Paper presented in Session No.97, "Implementing Geoinformatics for knowledge Integration and Decision Management:, 2002 Denver Annual Meting (October 27-30, 2002) http://gsa.confex.com/gsa/2002Am/finalprogram/index.html, Colorado.

Verhagen J., H. Booltink and J. Bouma 1995. Site specific management: balancing production and environmental requirements at farm level. *Agricultural systems* 49,369-384.

Wang, Y., and Fang, K.T. 1981. A note on uniform distribution and experimental deisgn. *KeXue TongBao* 26, 485-489.

Yu, P. L. 1985. *Multiple criteria decision making: concepts, techniques and extensions.* Plenum, New York.

Zeleny, M. 1982. *Multiple criteria decision making.* McGraw-Hill, New York.

Some Important URLs

www.gisdevelopment.net

http://www.geog.ubc.ca/courses/klink/gis.notes/ncgia/u59.html, 1991.

http://www.amesremote.com/section4.htm

http://journal.info.unlp.edu.ar/journal/journal8/papers/JCST-Apr03-7.pdf

http://gis.esri.com/library/userconf/proc98/PROCEED/ABSTRACT/A560.HTM

❑❑❑

GEOINFORMATICS APPLICATIONS IN AGRICULTURE, 2007
Eds. : A.K. Singh and U.K. Chopra; pp. 79-102
New India Publishing Agency, New Delhi (India)
E-mail : newindiapublishingagency@gmail.com
Web: www.bookfactoryindia.com

3

Geoinformatics Applications in India-Retrospect and Prospects Experience of National Remote Sensing Agency (NRSA)

P.S. Roy

Key words : NRSA, Natural Resource and Disaster Management, Agricultural Drought, Ecosystems, Biodiversity Information System, Water Resources, Sea Surface Temperature.

Introduction

A few decades ago, paper maps were the principal means to synthesize and represent geographic information. Manipulating this information was limited to a manual, non-interactive process. Since then the rapid development of new technologies to collect and digitize geographic data, together with an increasing demand for both interactive manipulation and analysis of this data, has generated a need for dedicated software's namely geographic information systems (GISs).

A GIS is more than a cartographic tool to produce maps. It stores geographic data, retrieves and combines this data to create new representations of geographic space, provides tools for spatial analysis, and performs simulations to help expert users to organize their work in many areas like public administration, transportation networks, military applications and environmental information systems. These above mentioned are few among many others where we are concerned with geographical locations, their present status, its potential for developmental activities, patterns of change over the years and the processes on the surface of the earth.

Geographic information technology is used to manipulate objects in geographic space, and to acquire knowledge from spatial facts. Spatial information theory provides a basis for GIS by bringing together fields that deal with spatial reasoning, the representation of space, and human understanding of space.

Spatial reasoning addresses the inference of spatial information from spatial facts. It deals with the framework and models for space and time, and the relationships that can be identified between objects in a spatio-temporal model of real world phenomena.

Scientific methods for the *representation of space* are important for the development of data models and data structures to represent objects in spatial databases. Spatial databases are distinguished from standard databases by their capability to store and manage data with an extent in space and time (spatial data types).

The *human understanding of space,* influenced by language and culture, plays an important role in how people design and use GIS.

The theory is used for the design of high-level models of spatial phenomena and processes. They are then mapped into conceptual, logical and physical models of spatial databases. It is the database that holds the data; without it, no useful function can be performed. Data are entered into the database in input processes. Later, they are extracted from the database for spatial analysis and display. The processes of data management, analysis and display are often supported by rules that are derived from domain experts. Systems that apply stored rules to arrive at conclusions, are called rule-based or knowledge-based systems. Those that support decision making for space-related problems are known as spatial decision support systems (or SDSS). They are becoming increasingly popular in planning agencies and management of natural resources.

All these activities happen in a social, economic and legal context. It is generally referred to as spatial information infrastructure. Everything within this infrastructure and all concepts of space and time in turn are shaped and determined by the cultural background of the individuals and organizations involved.

Evolution of Geoinformatics

The discipline that deals with all aspects of spatial data handling is called Geoinformatics. It is defined as *the integration of different disciplines dealing with spatial information.* Geoinformatics has also been described as "the science and technology dealing with the structure and character of spatial information, its capture, its classification and qualification, its storage, processing, portrayal and dissemination, including the infrastructure necessary to secure optimal use of this information" (Groot, 1989). Ehlers and Amer (1991) define it as "the art, science or technology dealing with the acquisition, storage, processing production, presentation and dissemination of geoinformation." A related term that is sometimes used synonymously with geoinformatics is geomatics. It was originally introduced in Canada, and became very popular in French speaking countries. The term geomatics, however, was never fully accepted in the United States where the term geographical information science is preferred. Goodchild (1992) defines GIS research as "research on the generic issues that surround the use of GIS technology, impede its successful implementation, or emerge from an understanding of its potential capabilities."

National Remote Sensing Agency (NRSA) and Geoinformatics

NRSA being the pioneer organization in remote sensing technology and applications, some of its experiences are highlighted in the following sections.

Aerial surveys using a variety of sensors, namely infrared scanners, multispectral scanners and radiometers etc. for monitoring crops such as coconut plantation in Kerala, sugarcane in Madhya Pradesh, for evaluating crop yield, has been the major milestone in the applications of remote sensing in India. Extensive aerial surveys supported by ground truth observations and measurements in a few selected test sites, for mapping soil salinity and/alkalinity, land use/land cover,urbanisation and water pollution studies carried out subsequently, have provided a sound base for remote sensing applications programme in India. Realising the application potential of space technology in general, and remote sensing technology in particular, the Indian Space Program was formulated in 1960's. The launch of Aryabhatta in 1975 and Bhaskara-I and II with a two band TV camera system - one in the visible and the other in the near infrared, along-with a three frequency. Satellite Microwave Radiometer (SAMIR) in 1979 and 1991, respectively using Soviet carrier, has provided a sound base for design and development of operational polar orbiting satellites. The TV payload with 1 km spatial resolution was designed for land applications whereas the SAMIR with a footprint of 125 Km for oceanographic / atmospheric applications. The establishment of satellite data reception facilities at Shadnagar, Andhra Pradesh by the National Remote Sensing Agency, Hyderabad was a major step forward in this endeavour which has provided the scientists from Department of Space and other user agencies, an opportunity to gain experience in the interpretation and / analysis of spaceborne multispectral data.

Several Joint Experiment Projects (JEP) were subsequently launched with the active involvement of the user community for management of natural resources. Subsequently, the National Natural Resources Management System (NNRMS) was established to ensure most effective utilization of remote sensing technology by facilitating the establishment of necessary space, ground and user segments, and integrating the views of the administrators, decision makers and technologists and user community. Concerted effort in the design and development of spacecrafts and sensors led to the successful launch of the Indian Remote Sensing Satellite Sensors (LISS-I and II) having four spectral bands which are almost similar to first four bands of Landsat TM and 72.5 m and 36.25 m spatial resolution, respectively. Subsequently, IRS-1B and IRS-P2 were launched in 1991 and 1994 respectively.

Convinced with the potential of IRS-1A/1B LISS-I and II and contemporary satellite (Landsat MSS / TM and SPOT HRV-MLA/PLA) data, for regional and district - level management of natural resources and the need to develop sensors capable of addressing certain issues not addressed by the satellites operating during late eighties and early nineties, the IRS-1C with a unique combination of State-of-art sensors namely Wide Field Sensor (WiFS), Linear Imaging Self scanning Sensor (LISS-III) and Panchromatic (PAN) Sensor with stereo capability, providing information on natural resources at regional and / micro level, was launched in late 1995. With the establishment of operational capabilities of Polar Satellite Launch Vehicle (PSLV) through successful launch of IRS-P2 and IRS-P3 experimental satellites therefrom, India launched another remote sensing satellite IRS-1D, with the sensors identical to that of IRS-1C, in 1997. India launched the indigenously built state-of-the-art satellite called RESOURCESAT (IRS-P6) with multi spatial, spectral and temporal resolutions in October 2003 using PSLV.

Natural Resources & Disaster Management

All are aware that GIS applications are enormous. GIS is capable to identify, locate, perform change studies, pattern analysis and modeling for Natural Resources Management such as action plan for integrated sustainable development, locating waste disposal sites and natural hazardous sites and identity and analyze minerals, coal, timber and water resources. It is also being used for landslide hazard zonation, forest fires and earthquakes. GIS is also a tool for decision-makers for land and water resources assessment, development and management.

Utilization of space borne multispectral data from IRS series of satellites as well as foreign satellites for natural resources and disaster management are briefly discussed below.

Crops

The works carried out so far, in India, has demonstrated the utility of single date satellite data in the estimation of preharvest acreages of major crops, particularly in the single crop-dominated regions, with 90/90 accuracy / confidence criterion. Under the Crop Acreage and Production Estimation (CAPE) project, the preharvest acreage estimates of major crops viz, paddy, wheat, sorghum, ground nut, rapeseed, mustard, cotton and mesta / jute occupying about 80% of the cropped area, are being regularly carried out. The highs patial resolution (23.5m) IRS-1C/1D LISS-III data enabled identification of commercially important crops, viz. chillies, tobacco etc. which are grown under multiple crop situations.

Many horticultural crops *viz.* mango, coconut, arecanut, oranges and banana could be identified and their acreages estimated using high spatial resolution LISS-II and LISS-III data. Efforts need to be made to provide the production estimates for various horticultural crops. In addition, detection, acreage and production estimation of horticultural crops occurring in different storeys in as three-storey vegetation system need to be attempted using multispectral data from currently as well as planned earth observation satellites.

Another important element in crop studies, wherein remote sensing has contributed significantly, is the yield modeling. Statistical, meterological and / or spectral models are used for crop yield estimation. Within spectral models,two approaches viz., single date data spectral index and multi-date data spectral index-growth profile are in vogue. The single date data spectral index approach relies solely upon the data acquisition within a narrow critical period of maximum vegetation growth phase while multi-date approach warrants spectral data at different stages of crop growth within the season. However, availability of cloud-free optical sensor data is a major problem especially during the *Kharif* Season. With the constellation of multiple satellites *viz.*, IRS-1C/12D and the envisaged future satellites, the interval in the revisiting period of satellites is expected to further narrow down, enhancing thereby the probability of obtaining cloud-free data. In addition, attempts are also underway to incorporate the spectral information in the process-based on models and crop simulation models to improve the predictive capabilities of the remote sensing-based crop production estimations.

Satellite remote sensing has played an important role in the pest management. It has helped surveying and identifying the inaccessible large desert regions, where locusts breed and ovulate, thereby enabled the farmers / district administration to develop strategies for preventing desert locusts from spreading and for taking up appropriate measures for their control.

Agricultural Drought Assessment

Monitoring the crop conditions at regular intervals during the crop growth cycle is essential to take appropriate curative measures and to assess the probable loss in production. The remote sensing data can provide information on the occurrence and the areal extent of crop stress. However, the identification of the cause-whether by pests and diseases or by moisture stress or any other factor still remains a major limitation. When stress is caused by more than one factor, the situation becomes even more complex. Using remote sensing techniques in conjunction with supplementary information from weather observations or knowledge on localized deficiency / toxicitly of nutrients or pollutant elements, by deductive process one can still arrive at the cause of the stress.

Realising the potential of satellite-derived vegetation index (VI) which is sensitive to vegetation stress and serves as surrogate measure to assess agricultural drought, a nation-wide project titled "National Ahgricultural Drought Assessment and Monitoring System (NADAMS)' was launched in 1987 to monitor the drought during *kharif* (South-West monsoon) season which is agriculturally more important and is also rain dependent, by generating Normalised Difference Vegegation Index (NDVI) from temporal NOAA-AVHRR data. For implementation of preventive and / corrective measures for combating drought at mandal / block (an administrative unit)_level, ecomprehensive drought assessment involving integration of information on cropping pattern, social types, coherent rainfall, zone, with improved spatial accuracy is required. Advanced WiFS with less than 100m resolutions which is likely to be available from IRS-P6 mission, may enable generating such information.

Soils

Landsat-MASS/TM, IRS-1A/1B LISS-I and II data have been used to generate soil resources maps at scales ranging from 1:250,000 with the abstraction level of sub-groups/ association thereof and association of families, respectively. In addition, the IRS-1C PAN stereo data have been used for generating information on soil resources at (1:5,000) maps with abstraction level of the phases of soil series and / individual series, could be generated from the proposedCartosat-1 (IRS P-5) PAN Stereo data with 2.5m spatial resolution.

Further more, derivative information, namely land capability, land irrigability, erodibility, reclamability, and suitability, for different crops which in turn enable preparing the optimal land use plan and in taking up land reclamation measures, wherever required, have also been generated. Models for soil moisture estimation using single frequ3ency, polarization and look angle SAR data, have also been developed. An operational model for estimation of both surface as well as root zone soil moisture using multi-frequency, multi-look angle and multi-polarization, microwave data available from future missions needs to be developed. A prototype expert system for soil classification using remote sensing data has also been developed.

Models for land evaluation could be developed in a GIS environment for automatic generation of derivative maps. Efforts need to be made to develop a remote sensing based operational expert system for soil classification in association with concerned national organizations.

The hyper-spectral measurements may be made covering different types of soils to identify the diagnostic spectral bands for detection and identification of soil, which may ultimately lead to the development of a dedicated sensor for soil resources mapping.

Degraded Lands

Based on the state-of-art of space borne multispectral data, the information on the extent, spatial distribution and magnitude of eroded lands, salt-affected soils, waterlogged areas, shifting cultivation, to name a few, at 1:250,000 scale and at 1:12,500 scale has been generated. This information has been used for planning land reclamation and soil conservation programmes. In addition, the monitoring of above-mentioned lands could also be carried out at regular interval.

For on –farm development, the assessment of water logging and soil salinity and/alkalinity is required at scales larger than 1:5,000 which could be generated through conjunctive use of a PAN and LISS-IV data available from the proposed Cartosat-1 and IRS P-6 missions. Efforts are needed to detect soil salinity and /alkalinity, and water logging in the black soil region using hyper-spectral measurements. Water logging due to rising ground water table is currently not amenable to detection by remote sensing. Measurements made from thermal sensors and ground-penetrating radar (GPR) may provide the desired information.

Hitherto, only qualitative assessment of soil erosion by water has been made with the existing satellite data which is relatively subjective. Soil loss from a watershed need to be quantified using empirical models, namely, Revised Universal Soil Loss Equation (RUSLE) or Water Erosion Prediction Project (WEPP). The values for management factor (C) and conservation practice factor (P) could be derived from fine resolution PAN stereo and multispectral data from IRS-1C / 1D and the state-of-the-art RESOURCESAT satellites. Further improvement in the level of such information is feasible from better spatial resolution data which is likely to be available from IRS-P5 (Cartosat–1), the future Indian mission.

Forests and Ecosystems

Satellite remote sensing has enabled generation of forest cover maps and to monitor changes therein at operational level. In India, the management of forest resources presently involves a three-tier monitoring system. At the national level, biennially forest cover mappings is carried out using satellite data by the Forest Survey of India (FSI) on 1:250,000 scale, and the relevant inputs required for forest management at the state level are generated at 1:50,000 scale or even at 1:15,000 scale. The advent of high resolution IRS-1C/1D L-III and PAN data has enhanced the capability to prepare forest type and large-scale stock maps. The complementary role of PAN data to that of aerial photography has substantially reduced the cost factor. The grassland mapping is yet another area where remote sensing has played a key role.

Remote Sensing holds substantial promise for providing reliable and timely information on potential, extent and composition of various eco-systems. In addition, the rates of afforestation / deforestation could also be assessed using multi-temporal satellite data. Currently, the forest fire damage assessment and the detection of active fire scars are amenable to remote sensing techniques. The assessment of forest damage due to minimizing activities also could be made using temporal multispectral data.

Efforts need to be concerted on monitoring the atmospheric carbon dioxide resulting from afforestation/deforestation as well as forest fires using data from the proposed IRS-2 (ATMOS-1) and EOS-CHEM. The forest fire risk models could be developed by integrating the information on topography, climate and *in situ* observation with the satellite-derived vegetation maps in a GIS domain. Besides, the environmental management plan (EMP) could be developed by integrating the information on land, water and atmosphere in a GIS environment. The assessment models monitoring urbanization and heat island formation and study of biogeochemical cycles may be attempted using measurements / observations made by future missions.

Biodiversity Characterisation at Landscape Level

Biodiversity characterisation at landscape level is attempted in North Eastern region, western Himalayan region, Western Ghats regions of the country using satellite remote sensing and GIS. These regions due to unique ecological setting are known for their unique flora and fauna their high endemism and rarity. The study is particularly important from scientific point of view as large diverse landscapes have been analysed at spatial extent and quite significant results on fragmentation, disturbance regimes and biological richness were observed. Findings revealed that high fragmented forests have fewer plant species, reduced anthropogenic use evidence and newer community types than medium and low fragment forests. It may also be concluded that slow fragmentation produces a decreasing trend in plant species diversity as revealed from the study. The landscape perspective is fundamental to understanding the role of disturbance. Disturbance regimes vary across landscapes as a function of topography and substrate.

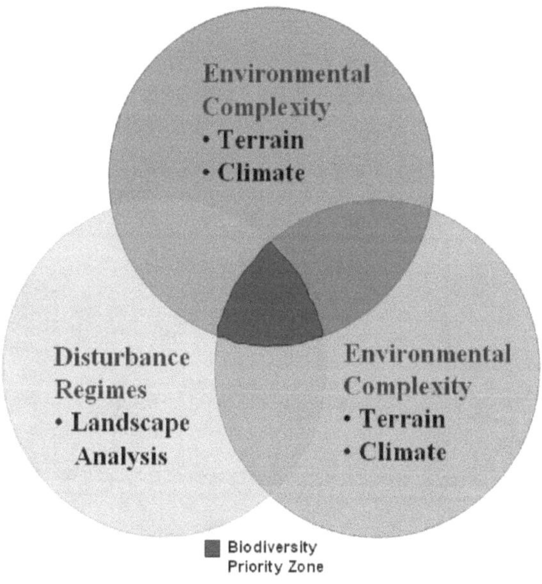

Biodiversity priority zone

Many of the natural ecosystems have become smaller and their surroundings have been modified. The biological richness is related to higher species diversity, higher degree of ecosystem uniqueness, higher degree of terrain complexity and lower levels of disturbance. Another dimension, socio-economic value of each species was also considered for biological richness. The degraded and barren areas were found to be least biologically rich. The biological richness maps very well reflect the distribution and status of biodiversity in the region. The information generated in this study would be of immense use if linked to other planning and policymaking that directs resource allocation, proper land use planning and sustainable consumption of natural resources.

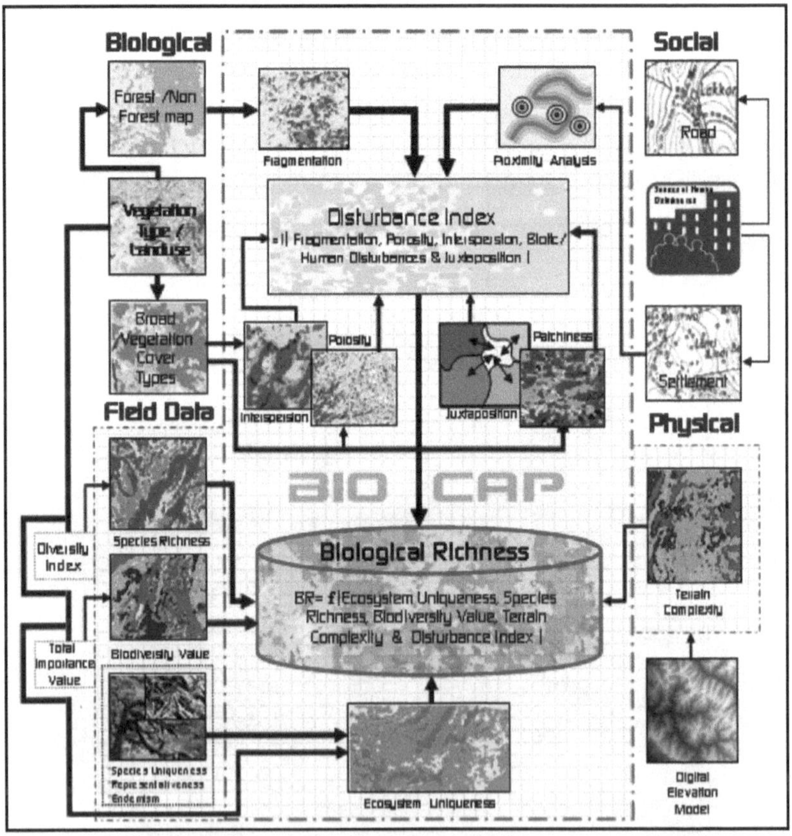

Schematic representation of Biological Richness Mapping at Landscape Level

Biodiversity Information System (BIS)

The motive of conceptualization of BIS is collection and organization of the available but distributed spatial and non-spatial database, into an interactive system, which is capable of presenting a user friendly and easily accessible interface to the remote users by using internet domain. One of the main components of the Biodiversity Information System is BIOSPEC query shell; the main objective of the BIOSPEC is to establish linkages between spatial and non-spatial database of networking Institutions for Bioprospecting and biodiversity conservation, prioritization etc. Presently the, existing Information system is available in the form of Website (www.bisindia.org). The website is in windows platform where the ASP technology is used for front end designingOracle9i data server is used for database management and spatial database engine ArcSDE is used for linking spatial and non spatial database .We are using Internet map server ArcIMS 4.0 for launching the spatial data on internet by using JAVA servlets. It is for this reason the Biodiversity Information System was conceptualized as a 5-component unit. The components are described on page 92.

Land Use/Land Cover

Space-borne multispectral data have been used to generate land use /land cover maps, urban land use and sprawl maps and wasteland maps at operational level. Nationwide land use/ land cover mapping on 1:250,000 scale using IRS 1A /1B LISS I data has been carried out. The IRS-1C/1D WiFS data offer an unique opportunity to generate the land use/ land cover maps at regional levels. Merging high spatial resolution IRS-1C / 1D PAN data with multi-spectral LISS III data provides adequate details for town planning. Wasteland maps on 1:50,000 have provided a sound data base for agro-climatic regional planning for increasing food production, whereas information on wastelands helped reclamation of degraded lands respectively. In addition, nation-wide wetland mapping has been carried out using IRS LISS–II data for comprehensive planning of wetland ecology and inland fisheries.

BIS Home page and its components (www.bisindia.org)

The future high resolution satellites have a potential for (i) Development of Spatial Decision Support System (SDSS) for land use classification (ii) Wasteland monitoring at larger scale, and (iii) Development of Land Information System (LIS) and Urban Information System (UIS); and urban / rural resources conflict models, among many other uses.

Water Resources

The Current satellite remote sensing capabilities for irrigation water management include end-of-season evaluation of canal command areas at the disaggregated level and diagnostic analysis of problem distributaries to enable follow - up corrective management. For through-the-season management, IRS-1C /1D WiFS data with a repetivity of 5 days in conjunction with multi frequency, multi-look angel and multi-polarization SAR microwave data could be used. The detection of physical degradation such as canal seepage and excessive sedimentation or weed growth in canals, through high resolutions panchromatic data from IRS-1C/1D/P5 supported by LISS-III / LISS -IV multispectral data from IRS P-6, may be attempted.

The information on near-realtime flood inundation, and damage assessment for monitoring flood events, post-flood river configuration to assess vulnerability of flood control structures, preliminary flood hazard risk zone mapping and flood forecasting could be generated using currently available satellite data. However, comprehensive flood monitoring would call for integration of ground measurements and aerial / satellite remote sensing data along with the GIS on the flood plain characteristics including topography in a GIS domain. The flood risk zone maps need to be refined with high spatial resolution data and digital elevation models based on close contour information. Furthermore, the multi-temporal satellite data have been used for assessment of water spread in the reservoir.

Currently, seasonal snowmelt forecast is being given for sutlej river and weekly forecasts for Beas and Parbati basins.

There is a need to develop improved snow-melt run-off models using multi-frequency, multi-look angle and multi-polarization SAR data for deriving information on snow cover characteristics.

Using Landsat - TM, IRS 1A / 1B LISS II data maps at 1:250.000 scale and 1:50,000 scale the extent and spatial distribution of glaciers have been generated. Further improvement could be made using IRS 1C / 1D LISS III and PAN and proposed Cartosat 1 PAN and IRS P-6 and LISS IV merged data. Efforts need to be made to fine tune various hydrologic models, namely, rainfall run off prediction using multitemporal high spatial resolution space borne multispectral data.

Geosciences

With the currently available space borne multispectral data, the geological, geomorphological and hydrogeomorphological maps on 1:250,000 scale have been prepared at operational level. The National Drinking Water Technology Mission Project encompassing generation of hydrogeomorphological maps for entire country at 1:250,000 scale is an example of such capability. However, for ten states such maps have also been prepared at 1:50,000 scale. By merging LISS III and PAN data, detailed maps up to 1:15,000 scale also covering smaller area are being generated. Besides, targeting the areas for mineral exploration as well as the detection of thermal anomalies arising out of volcanic eruption and underground coalfire, have also been undertaken. The IRS-1C LISS III and PAN merge data enabled detection of areas for diamond exploration. In addition to existing techniques, stress need be given on new methods incorporating computer based mathematical models, particularly for concealed deposits.

The aeromagnetic observations are powerful tools for geological mapping and mineral exploration. The integration of aeromagnetic data with remote sensed data has not only helped in delineation of sub surface basement structures controlling the occurrence of hydrocarbon-bearing sediments, but also in the estimation of the thickness of sediments in the Cauvery and Kutch basins. Conjunctive use of multi-spectral thermal data, satellite altimetry, radar,a nd aeromagnetic data may be made for better prospecting of oil field detection.

The hyperspectral measurements in the visible / NIR / SWIR / TIR region with a spectral resolution of 10 nm have been found to be quite useful for identification of different rock forming minerals. The geological application potential of such a sensor has to be explored in detail in the Indian context. Apart from such studies, operational models for ground water yield may be developed from information on ground water recharge, draft and balance derived from remote sensing data.

The information on phase difference obtained from microwave signals can be effectively used to detect the micro-relief changes resulting from earthquakes and landslides, and other dynamic phenomenon which is an important area for taking up pilot studies. In addition, landslide-prone areas could be delineated using higher spatial resolution data with stereo coverage from IRS-1C / 1D and future IRS-P5 missions. The GPS network may be useful in monitoring tectonic movements.

Integrated Surveys

For sustainable development of any region, the information on available land and water resources need to be suitably integrated with other collateral and socio-economic data in GIS domain to arrive at locale-specific prescriptions on a watershed basis. The locale-specific action plans, to illustrate, include water harvesting structures, soil conservation through terracing and contour bunding, afforestation, agro-forestry and agro-horticulture, and fuelwood and fodder development. Currently, the IRS-1B LISS II data are being used for generating information on land and water resources at 1:50,000 scale under Integrated Mission for Sustainable Development resources at 1:50,000 scale under Integrated Mission for Sustainable Development (IMSD) project. The IRS-1C PAN data are also being also used for identification of individual fields where a specific action plan is to be implemented by digitally superimposing the cadastral map boundaries over digital PAN data.

For generating action plan at village/micro watershed level, cadastral-level maps for land and water resources need to be prepared using proposed Cartosat-1 PAN stereo data with 2.5 m spatial and better than 5 m Z-axis resolution in conjunction with the multispectral LISS-IV data with better than 10m spatial resolution available from future IRS-P6 mission and integrating in GIS domain. The IRS-1C LISS-III data are currently used for impact assessment of the implementation of action plan. It could be further improved with the availability of higher spatial resolution multispectral data LISS IV from IRS P6. The National (Natural) Resources Information System (NRIS) with the digital database on natural resources and related cadastral information is being established at district / taluk headquarters to enable decision makers and planners to use this information for planning and implementation of locale-specific prescriptions/action plan at gross-root level.

The rule-based expert system for objective integration of information on land and water resources, and socio-economic data need to be developed for generation of action plan. It could be achieved by developing a general model and subsequently incorporating the values for locale-specific variables to suit the local conditions.

Ocean Applications

The sea surface temperature (SST) and potential fishing zone (PFZ) maps have been generated routinely using NOAA-AVHRR data. Currently, the PFZ maps are generated bi-weekly on every Monday and Thursday using ¾ days SST composite. These PFZ maps are of very coarse resolution. The PFZ maps have helped fishermen to get more fishes. The catch per unit effort (CPUE) is more wherever PFZ maps have been used as compared to situations where it was not done so. To be more effective, high resolution PFZ maps showing the maximum details for easy approach need to be generated. The integration of Sea Surface Temperature (SST) with chlorophyll concentration is required for PFZ forecast which could be achieved using the current and future planned ocean colour missions like Sea WiFS (SeaStar), ADEOS-OCTS and OCM aboard OCEANSAT-1 and 2. Using IRS P3 MOS-B data, attempts have been made to retrieve chlorophyll, suspended sediments, and yellow substance and aerosal optical thickness using linear inverse PCA model. The estimation of columner primary productivity using spectral and non-spectral models to predict the mixed layer production for fisheries forecasting and carbon assimilation studies in the oceans are in progress.

The information on altitude of the satellite orbit provided by altimeter aboard current and future missions, may help studying the deviation in the ocean geoids. Microwave sensors play an important role in monitoring the spread of oil slicks. An operational methodology for identification and estimation of its precise thickness and spread need to be perfected.

Sea topography variations derived from altimeter, sea surface winds from scatterometer, sea-surface temperature from AVHRR/ATSR and salinity from L-band microwave radiometer are some of the inputs to ocean circulation prediction models-both wind-driven and thermohaline types. The altimeter and Ocean Colour Monitor (OCM) data from the proposed OCEANSAT could be used for generating information on wave energy and directional spectrum, is essential and extremely useful for coastal zone development and execution of off-shore projects. Coastal zone mapping for entire Indian coast has been carried out using IRS-1B LISS-I/II data which has enabled defining coastal regulation zone to protect coast line from exploitation and helped generating optimal land use planning. In addition, coral reef mapping was also carried out using IRS-1A/1B/1C LISS-II / LISS-III data which has enabled identifying and ultimately protecting areas of active coral reef and helped earmarking some areas as marine parks.

Atmospheric Applications

The information on various atmospheric parameters, namely temperature and humidity profile, wind, atmospheric pressure, sea surface temperature, ocean circulation, swells, etc. are extremely important for weather forecasting. Currently, information on cloudiness and winds derived from INSAT data and sea surface temperature and humidity profile from NOAA data are being used for weather forecasting. Improved forecasting seems feasible from improved / additional information on surface pressure, cloud height, precipitation, and sea surface and mid-level winds derived from future space missions. The temperature profiles from the surface to 10 mbar level, water vapour content, and ozone content derived from TOVS-TIROS operational vertical sounder could be used for numerical weather prediction (NWP). Timely and reliable information on cyclone development and movement is a pre-requisite for taking up precautionary measures for combating disaster as it results in saving several human lives and property. The INSAT-VHRR / DWS is currently used for disaster warning which is, by and large, satisfactory. Efforts are, however, on for further improvement in the efficiency and precision of existing disaster warning system.

The EOS-CHEM-Ozone Dynamics Ultraviolet Spectrometer (ODUS) and EOS-AM-1 and Measurement of Pollution in the Troposphere (MOPITT) may provide information on air pollution whereas MOS aboard IRS-P3 may provide the desired information on large-scale air pollutants and the upwelling zones.

The satellite remote sensing has been found to be very useful in the assessment of spatial distribution and concentration of ozone. The sensors aboard IRS-2 series of satellites, Total Ozone Mapping Spectrometer (TOMS) and Ozone Dynamics Ultraviolet Spectrometer (ODUS) aboard EOS-CHEM series may enable regular monitoring of global ozone.

Climate / Geosphere and Biosphere Programme

Realising the importance of spacedborne measurements in providing information on climate change due to anthropogenic disturbances, the Indian Space Research Organization (ISRO) has initiated a programme titled "ISRP Geosphere and Biosphere Programme". The programme includes around 40 research projects dealing with each core theme of the International Geosphere and Biosphere Programme (IGBP). The salient features of the programme includes the development of digital database on land use / land cover, India evapotranspiration modeling, development of biophysical algorithms, studies on trace gas fluxes and development of models for biomass and net primary productivity using IRS data. Future plans include development of soil-vegetation-atmosphere model using spaceborne multispectral measurements in conjunction with in situ observations.

Cartography

The space borne multispectral data have been extensively used for preparing thematic maps at scales ranging from 1:250,000 to 1:50,000 using Survey of India topographical sheets as base map. The IRS-1C/1D PAN data with a spatial resolution of 5.8 m have been found adequate enough for identification of features for 1:50,000 scale mapping. The updation of topographical sheets at 1:250,000 scale and subsequent compilation to 1:50,000 scale, and the generation of Digital Elevation Model (DEM) have been achieved using IRS 1C/1D, LISS III and PAN merged data and ground control points (GCP) identified by Global Positioning System (GPS). In addition, contour maps with perspective views have been generated from aforesaid data base with 20m contour interval which meet the cartographic requirements of 1:50,000 topographical sheets. The improved spatial resolution data with stereo coverage from Cartosat-1 may provide the information necessary for preparation of 1:25,000 scale topographical sheets.

Digital Photogrammetry

With the state-of-the-art technology, satellite data have been used for preparing ortho-images and for town planning surveys. Besides, urban plan maps and town plan maps could also be generated from existing satellite data with 5.8m spatial resolution and stereo coverage. Besides, such data could be utilized for generating the Digital Elevation Model (DEM) for critical areas, like dam site, watersheds, etc. Taking advantage of the future developments in the sensor technology, efforts need to be made to generate the information required for route planning, sewage routing, preparing micro plans and guide maps; the assessment of urban flooding and its modeling through wire mesh leveling need to be attempted.

Conclusions

India has established the operational capability of launching earth observation satellites, data reception, processing, interpretation and / analysis for providing timely and reliable information on natural resources and environment to the users. The Indian space programme is user driven. Establishment of NNRMS involving user Ministries of the Govt. of India, is a major step forward in this direction. The availability of spaceborne multispectral data with improved spatial, spectral and temporal resolution from IRS 1C / 1D/ Resourcesat has further enhanced such capability. Efforts are on to utilize the satellite data in newer application areas and to refine the existing application capabilities vis-à-vis the user needs. In addition, advanced technologies like digital photogrammetry, Digital Elevation Modeling (DEM), Global Positioning System (GPS), artificial neural network (ANN), fuzzy logic, contextual and textural classification algorithms, internet based applications, geo-spatial data interoperability, web services are also being tried out for improving the quality and information content on natural resources and environment. Also plans are afoot to launch the dedicated missions for specific applications viz. Oceansat, Cartosat, RISAT etc., to cater to the future requirements of natural resources and environmental management. In addition, atmospheric applications which have not been addressed so far are planned to be included in the future space programme.

References

Bhattacharya, A. and Reddy P.R. 1991, Hydrogeomorphological mapping for ground water prospects in India using IRS imagery - How to meet the demand of drinking water. In Shunji Murai edited Remote Sensing in Asia and Oceania - Environmental Change and Monitoring Asian Association of Remote Sensing, Tokyo, Japan.

Chari, S.T. 1988, National agricultural drought assessment and monitoring systems in India: The satellite contribution. 39th IAF Congress, Special Publication on Space and Drought Assessment, ISRO - 44-89, Bangalore, India

Dodge, A.G. and Bryant, ES 1976, Forest type mapping with satellite data. *Journal of Forestry*, 74., 8 pp 526-531.

Dwivedi, RS 1992, Monitoring and the study of the effects of image scale on delineation of salt-affected soils in the Indo-Gangetic plain. *International Journal of Remote Sensing* 13(8). pp 1527-1536.

Ehlers, M. and Amer, S., 1994, Geoinformatics : An integrated approach to acquisition, Proceeding and Production of Geo-data. In Proceedings EGIS '91, Brussels pp 306-312, 1991.

Goodchild, Michael F., 1992. Geographical Information Science, Int. J. Geographical Information Systems, 6(1): 31-45.

Groot, Richard 1989 Meeting educational requirements in Geomatics, *ITC Journal*, 1989 (1): 1-4.

Jayaraman, V., Chandrasekhar, M.G., and Rao U.R. 1996, Managing the natural disasters from space technology inputs. Proc. 47th International Astronautical Congress, October 7-11, 1996/ Beizing, China.

Nagaraja, R., Gautam, N.C., Rao D.P. and Raghavaswamy V 1992, Application of Indian Remote Sensing Satellite (IRS) data for wasteland mapping. In R.L. Karale-edited Natural Resources Management: A new perspective. National Resources Management System (NNRMS) Dept. of Space, Bangalore pp 170-174.

Nagaraja, R, Narayanan L.R.A., Rao D.P. and Gautam N.C., 1989, Utility of Landsat TM data for Wasteland development, *The National Geographical Journal of India*, 35(1), pp 38-52.

Navalgund R.R., Parihar J.S., Ajai and Nageswar Rao P.P. 1991. Crop inventory using remotely sensed data. *Current Science* 61(3 & 4): 162-171.

Roy P.S., Joshi P.K., Singh S., Agrawal S. and Yadav D. 2002. Biome level characterization of Indian vegetation using IRS-1C WiFS data. Proceedings of the International Workshop on Asian Forest Cover Assessment & Conservation issues. 12-14 February 2002, IIRS, Dehradun, India. pp. 165-192.

Roy, P.S. 1994. Possibility of using spectral models for estimating growth conditions of Indian forests. *NNRMS Bulletin* 18: 26-27.

Roy, P.S. and Ravan, S. 1996. Biomass estimation using satellite remote sensing data - an investigation on possible approaches for natural forest. *Journal of Biosciences* 21(4): 535-561.

Roy, P.S., Sharma, K.P. and Jain, A. 1996. Stratification of forest density in dry deciduous forest using satellite remote sensing digital data - an approach based on spectral indices. *Journal of Biosciences* 21(5): 723-734

Roy P.S. and Tomar S., 2000. Biodiversity Characterisation at Landscape level using Geospatial Modelling Technique. *Biological Conservation* 95(1): 95-109.

Rao, D.P. and Chandrasekhar, M.G. 1996, Integrated Mission for Sustainable Development (IMSD): A holistic approach to land and water resources management. Presented at the 47th Inter-national Astronautical Federation Congress, Beijing, China, October 7-11, 1996.

❑❑❑

GEOINFORMATICS APPLICATIONS IN AGRICULTURE, 2007
Eds. : A.K. Singh and U.K. Chopra; pp. 103-120
New India Publishing Agency, New Delhi (India)
E-mail : newindiapublishingagency@gmail.com
Web: www.bookfactoryindia.com

4

Horticultural Resources Information System – A Geo-informatic Tool For Horticultural Development

S.C. Kotur, D. Sreenivasa Murthy, T.M. Gajanana, Reena Rosy Thomas, Chitra Mothkur, L.B. Ashok & G.S. Geetha

Key words : Horticultural Resource Information System. INARIS.

Introduction

Wide ranging Indian climatic conditions are well suited for boosting production of a variety of horticultural crops (fruits, vegetables, flower/ornamental/medicinal and aromatic crops) which presently hold an important position in the global scenario. Development was at a very low ebb till the Third Five-Year Plan and increased significantly since the Seventh Five Year Plan (Rs.10,000 millions plan outlay). Government of India established National Horticultural Board (NHB) in 1984 and Agriculture and Processed Food Products Export Development Authority (APEDA) in 1986 to facilitate its larger development. The outlay is expected to increase even more in the Tenth Plan. Significant progress has been made in area expansion of different horticultural crops resulting in higher production. Besides, use of modern technologies has also brought about improvement in productivity. More than 50 per cent increase in production is seen in many of the horticulture crops during last one decade.

The efficient planning and implementation of different developmental programs on horticultural crops depends on production parameters like soil suitability, varietals dominance, marketable surplus, marketing channels, post harvest handling practices, cost of cultivation and market availability for sustained production. In addition, information on latest technologies developed by various research organizations on horticultural improvement, production, protection and post-harvest technologies are needed to be collected and made available to the scientists, planners and managers working in the related fields. But the available information on this aspect is scattered and the database is too weak to provide adequate support for developmental strategies. The digitization of comprehensive database on horticultural crops will definitely enhance the pace of increasing the area, production, productivity and promoting industry of horticultural crops that have taken lead in the new millennium with reference to nutritional quality of food and export potential. A strong, well-knit database to support need-based, market-driven and cost-effective horticultural development is imperative. Therefore, under National Agricultural Technology Project (NATP) a Mission Mode Sub-project on "Integrated National Agricultural Resources Information System (INARIS)" has been taken up to evolve a comprehensive computerized data management system on various agricultural resources like crops, animals, fishes and natural resources by harnessing RDBMS and GIS. The Co-operating Center at Indian Institute of Horticultural Research, Bangalore has developed the "Horticultural Resources Information System" with the objectives of (i) integrating existing databases in horticulture, (ii) designing and developing a comprehensive database in horticulture and (iii) developing resource maps of major horticultural crops in the country by integrating attribute and spatial data through GIS. In addition to simple storage and retrieval it will include other functions such as manipulation and dissemination of information to various users. This database will be integrated with 13 other Co-operating Centers dealing with different agricultural resources through a virtual private network (VPN). The Lead Center at Indian Agricultural Statistics Research

Institute, New Delhi provide warehouse facility with a capability to analyze the data variously and provide report as cubes, tables and graphs to form a national database to cater to the needs of the users through a website. The facility of retrieving information as value-added maps using GIS will also be built in.

Methodology

Requirement Analysis for database development: A requirement analysis was carried out involving experts in horticulture and database management. A review of existing databases, identification of data needs of different clients, identification of data sources, data availability, storage level and data gaps were made. Data fields and coding structure were identified. Thematic maps to be generated using GIS were also identified. A detailed document was prepared by compiling the exercise done in all the Co-operating Center of INARIS (Sharma *et al.,* 2002).

Components of horticultural database: Four important components were identified (depicted in Fig 1) and a brief description on each of the component is given below.

Crop production and management: Complete information on production and management of crops like general information, variety details, and climatic requirement, production and protection details were included.

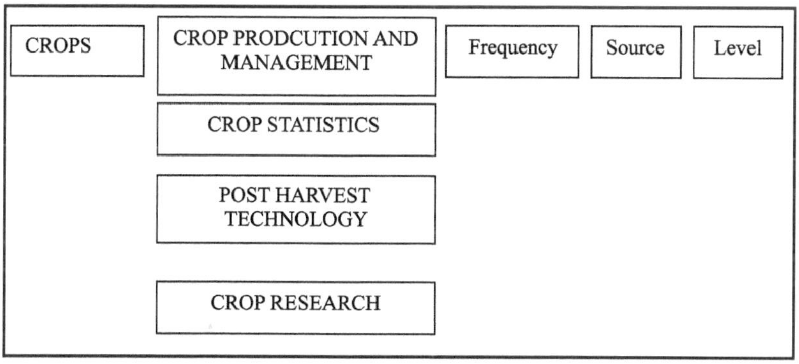

Figure 1. Data Structure of Horticulture Database

Crop statistics: Information on area, production, market prices and arrival, exports and imports were identified in the database.

Post-harvest management: Horticulture is the major contributor to processing like mango pulp; raisins, jams and database on this information was identified and included in the database.

Crop Research: Information on crop research regards to the projects in each of the research institutes of ICAR systems was included. Availability of comprehensive information will be useful to avoid the duplication of research efforts.

Crops in horticultural database: A large number of fruit, vegetable and other crops are grown in India. Considering the time and resources available during the project the crops as listed in Annexure I were included in the database for providing the information as well as thematic maps regarding the area of suitability, changes in production and area pattern in different horticultural crops.

Entities and relationships: The entities were identified and converted into logical records. Further, these were used to construct the physical records and horticultural database. The example of one such entity with attributes on general information is given in Fig 2.

Once all the entities were identified and defined, the relationships were established through different types.

Data structure: The basic data structure based on the components of horticulture database was constructed regarding the data fields, type, frequency, source, etc. The detailed data structure for the crop production and management in the horticultural crops is presented in Table 1. For illustration only a part of the data structure is presented.

Data dictionary: Software was developed at the Lead centre to create data dictionary and the data dictionary was developed. A report generated using data dictionary in horticultural crops is given in Table 2.

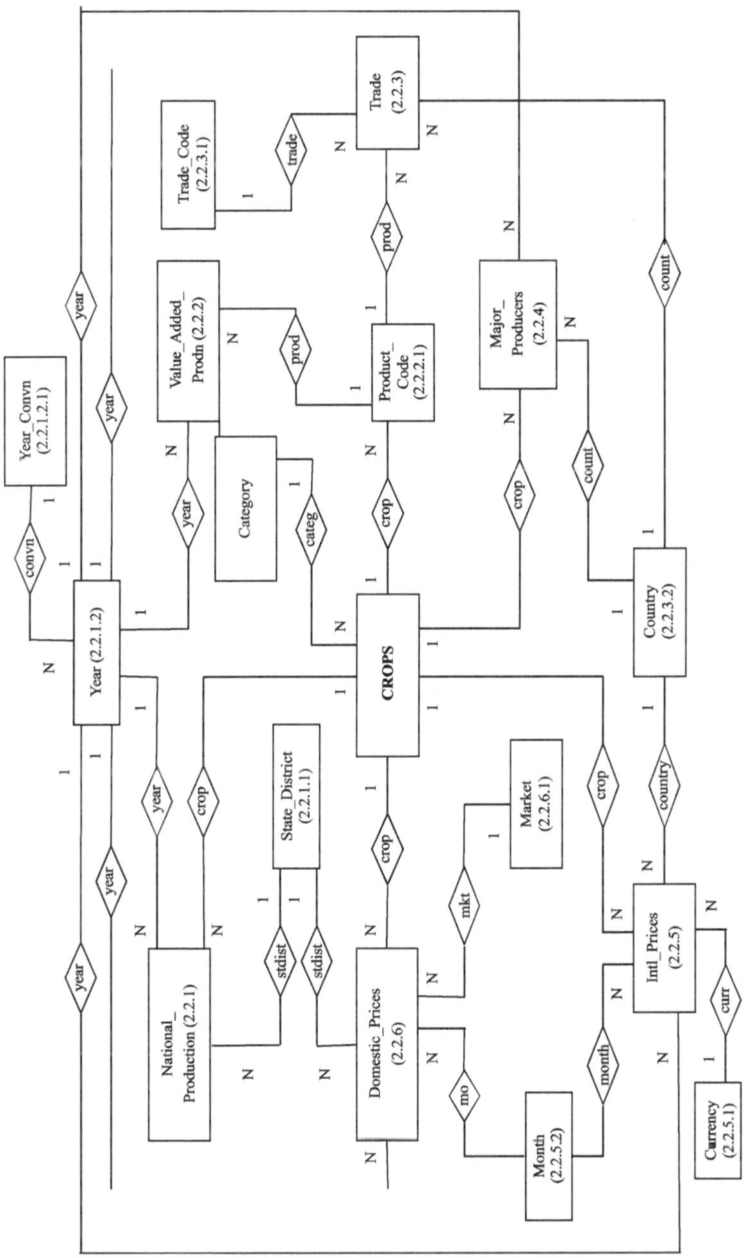

Figure 2. Entity-relationships (E-r) Diagram of Statistics Database of Horticultural Crops

Table 1. Data structure of Crop Production and Management in Horticultural Crops *

Item	Parameter	Data Fields	Data Type	Frequency of updating	Period of availability	Data Gaps	Storage Level	Source	Ge
General Information of the crop									
Name of the Crop	General Information	Crop_Sc_Name	Text	Permanent		Nil	National	IIHR/ICAR/SAU	
		Crop_Origin	Text						
		Crop_Genus	Text						
		Crop_Family	Text						
		Crop_Order	Text						
		Crop_Species	Text						
		Crop_Morphology	Text						
		Crop_Image	Text						
		Crop_Description	Text						
		Nutritive_Value	Text						
		Gen_Suitability	Text						
		Season	Text						
		Imp_Varieties	Text						
		Prodty_Range	Text						
		Producer_World	Text						
		Producer_India	Text						
		Other_Info	Text						
Information on varieties									
Name of the Crop	Varietal Details	Var_Code	Number				State/National	IIHR / SAU / Private	
		Var_Name	Text	Permanent					
		Var_Parent	Text						
		Var_Author	Text						
		Var_Purpose	Text						
		Var_Duration	Text						
		Var_Description	Text	Permanent					
		Var_Image	Image						
	Recommended Agency	Recom_Agency_Code	Number						
		Recom_Agency_Name	Text						
	AER	Recommended States / AER_Code	Number						
		AER Name	Text						
	Soil Requirement	Soil_Reqt_Code	Number						
		Soil_Reqt_Name	Text						
	Climate	Climate_Code	Number						
		Climate_Name	Text						
	Breeding Method	Breeding_Code	Number						
		Breeding_Name	Text						
	Other details	Recom_Year	Text						
		Pest_Resistance	Text						
		Disease_Resistance	Text						
	Season	Season_Code	Number	Permanent					
		Season_Name	Text						
	Yield details	Pot_Yield	Text						
		Avg_Econ_Yield	Text						
		Yr_First_Econ_yield	Text						
		Econ_Longitivity	Text						

* Only part of the data structure is presented

Table 2. Data Dictionary Report of Crop Production and Management in Horticultural Crops *

Field Name	Data Type	Size	Description
Crop_Code	Integer	10	Crop_Code
Crop_Sc_Name	Text	50	Crop_Sc_Name
Crop_Origin	Text	100	Crop_Origin
Crop_Genus	Text	20	Crop_Genus
Crop_Family	Text	20	Crop_Family
Nutritive_Value	Text	50	Nutritive_Value
Other_Info	Text	100	Other_Info
Crop_Order	Text	20	Crop_Order
Crop_Species	Text	20	Crop_Species
Crop_Image	Text	20	Crop_Image
Crop_Morph	Text	255	Crop_Morphology
Gen_Suitab	Text	50	Gen_Suitability
Imp_Variety	Text	50	Imp_Varieties
Season	Text	15	Season

* Only a part of the data dictionary is presented.

INARIS Data Dictionary Report
Crop Production and Management in Horticultural Crops
General Information:

Path analysis: This was carried out to verify both logical and physical designs and to test the correctness of their relationship. The database was then organized into a website using graphical user interphase (GUI).

Application of GIS: Database information will be utilized for developing the thematic maps on various components of horticultural database. While the software like Arc-GIS available to develop different thematic maps are very powerful the main hurdle seems to be the lack of well-defined spatial information in respect of crop production in terms of climatic, soil and crop requirement database. While efforts are on to collect the database a more concerted effort is required to collect and generate such geo-synchronized data before the fruits of GIS techniques can be savored in the process of development of horticulture.

Data

Data on various components were collected mostly from the secondary sources. Several as listed below are generating data in horticultural crops and many of these are published variably.

- Area, production and productivity of some important horticultural crops - published by the Directorate of Economics and Statistics, New Delhi
- Market information regarding the arrivals, process and the market details published by the National Horticultural Board, New Delhi
- Export and import information of horticultural crops by DGCIS, Calcutta
- Data on area, production and productivity of crops in their respective states published by State Departments of Horticulture.

In addition, many of the ICAR institutions like IIHR, Bangalore already published many expert system for the general use such as the Mushroom Expert System, Grape Expert System, SI HORT version 1.0 regarding the package of practices in horticultural crops in Southern India, Expert System on IPM in cabbage, serpentine leaf miner in tomato, etc. Such information was also made use to populate the database.

The maps from survey of India were used for generating the thematic maps using district as the mapping unit on various aspects of production of horticultural crops.

Results and Discussion

Development of horticultural database

Using Microsoft's Active Server Pages (ASP) with Microsoft FrontPage 2000 and planning to switch over to the Oracle backend in due course of time the front end of the horticultural information system (web pages) were developed. JSP (Java server pages), which is more advantageous than ASP was used to make the web pages dynamic. With JSP it is easier to move across platforms, and across web servers. Some of the web pages designed using ASP are presented in Fig 3 and 4. The home page of the Horticultural Information System is shown in Fig 3 while web page on components of Horticultural Information System is presented Fig 4. A user-friendly retrieval system is designed so that the user can retrieve information from the database regarding the horticultural crops without any difficulty. An output page indicating the area under chilli for Chittor district of Andhra Pradesh is shown in Fig 6. The databases on different agricultural resources in India will be integrated at Lead Centre, IASRI, New Delhi. The Lead Centre will provide warehouse facility to form a national database to cater to the needs of the users through a website poised to be launched soon The website has the capability to privide the information to the clients in the form of analyzed tables, cubes, graphs (histograms, line diagrams etc.,). There is a facility to retrieve different thematic maps generated using GIS. A bank of frequently asked questions (FAQs) is also being generated to provide usable information quickly to the farmers.

Thematic maps on change in area and production of Onion in India

Since INARIS aims at providing geo-synchronized information at district level on various crops, the relevant attribute information was collected for all the districts. The spatial database on soils is being collected at 1:1000,000 scales. Similarly climatic information is also being gathered. Integrating different databases and applying different decision criteria could develop thematic maps of different horticultural resources using GIS. Some of the themes where maps are useful are:

- Area, production and productivity of different horticultural crops/products during different years and at different temporal intervals.
- Marketing, trade and export of different horticultural commodities during different years and their temporal behaviour.

Figure 3. Home page of the Horticultural Information System

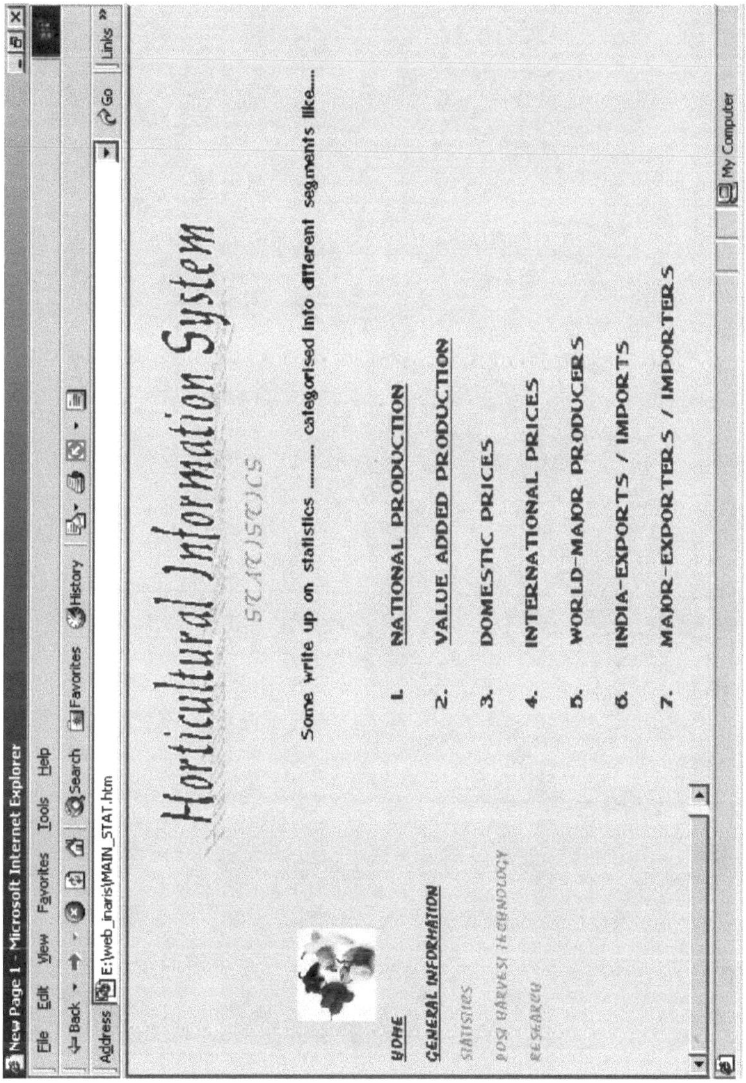

Figure 4. Components of Horticultural Information System

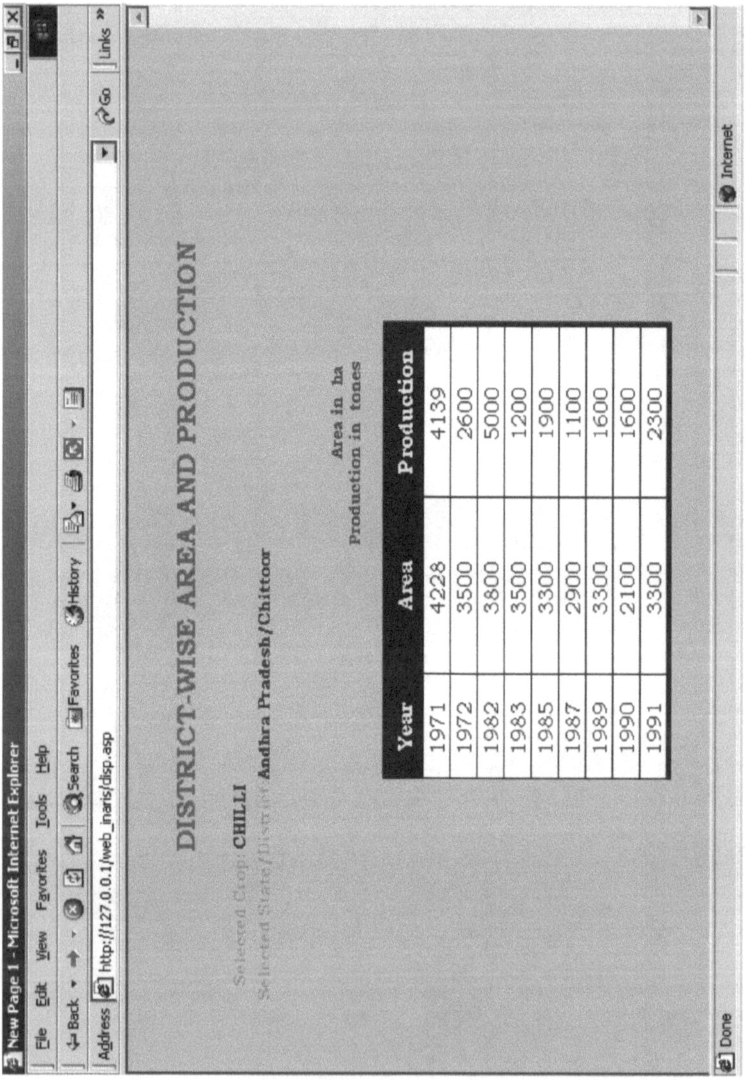

Figure 5. Output page of the horticultural System -

(iii) Suitability of different categories of lands for growing horticultural crops for use in planning future expansion of horticultural enterprise.

(iv) In well-organized production industries like grape, providing technological and logistic support on production, marketing and transport of different supplies and services.

The thematic maps on district-wise data on area and production of onion during 1981 and 1991 are presented in Figs 5 and 6. The salient changes that are evident from the GIS-generated maps are: (i) there was an increase of 18 per cent area under onion and 22 per cent in production between 1981 and 1991 and (ii) in majority of the districts the change was not discernible within the classes of area and production adopted to analyze the data. (iii) However, the area under onion increased during 1991 in 21 districts while decreased in 9 districts compared to that during 1981. (iv) In contrast the production showed distinct increase in 15 districts and reduction in 7 districts during 1991 relative to 1981.

It is possible to analyze the data for different purposes and generate thematic maps using GIS to evaluate the temporal trends and changes.

An evaluation of the database and retrieval of information from the Horticultural Resources Information System showed that:

- There are critical gaps in the information on area, production and other inputs in some years and for some districts which need to be filled/augmented.

- The spatial geo-synchronized database on climate, soils, land use, area and production under different crops is largely lacking. Different agencies like Central Research Institute for Dryland Agriculture (CRIDA) for climatic database, National Bureau of Soil Survey and Land Use Planning (NBSS&LUP) for soil database need to be worked out in a concerted manner to develop a uniform scale

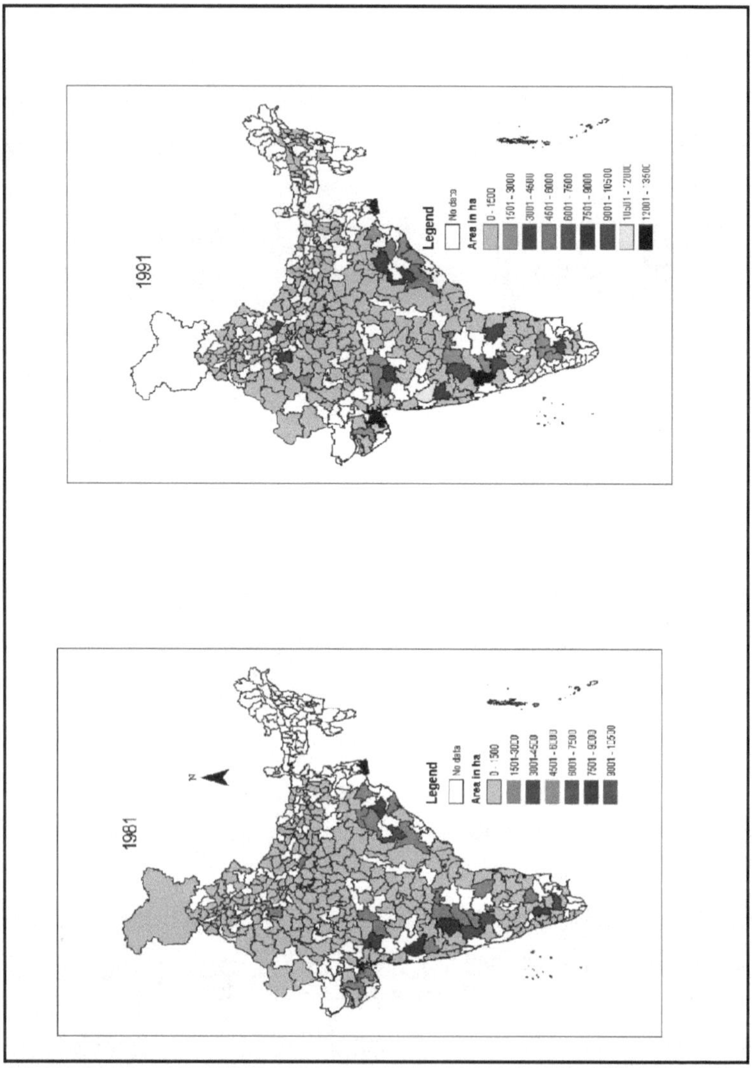

Figure 6. Thematic maps on district wise area of onion in India

comprehensive database to apply GIS software properly to derive different thematic maps. The aspect of agro-ecological zones also should be integrated into such databases.

- The decision criteria in terms of crop requirements of climate, soil and others need to be specified on agro-ecological and geological basis to develop different thematic maps.
- There is need to constantly monitor and update the databases to cater to the ever-increasing needs of the clients.

Future thrusts

There is scope and need to continually update the requirements of the clients, data and geo-synchronized inputs, FAQs and RDBMS/GIS/warehouse software application to make the information related to horticulture, available in the most user-friendly manner in consonance with the overall agricultural development. The scale requires enhancement from district to taluk/block and perhaps even at lower block levels. The information is at present in English and can also be made available in Hindi and other Indian languages.

Summary

To provide a comprehensive informational support to farmers, planners, bankers, scientists and students involved in the fast emerging development of horticulture, Horticultural Resources Information System (HRIS) was developed at IIHR, Banaglore as an integral part of Integrated National Agricultural Resources Information System (INARIS) under the aegis of the National Agricultural Technology Project (NATP). The requirement analysis was carried out in consultation with experts and scientists associated with horticulture to generate a blue print of HRIS. Data were collected from various secondary sources as well as from ICAR institute/SAUs and organized into a database using RDBMS. Using spatial data, different thematic maps were generated on different

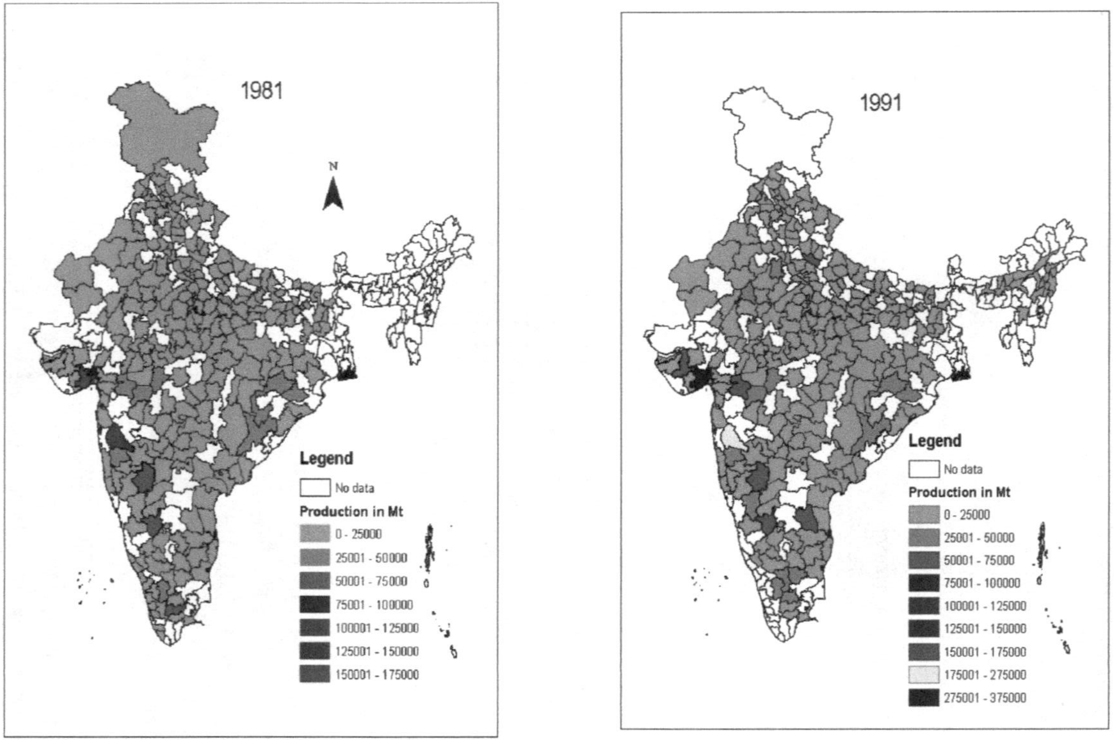

Figure 7. Thematic maps on district wise production of onion in India

horticultural resources. The HRIS was then integrated into the warehouse operated by the Lead Centre located at Indian Agricultural Statistical Research Institute. The HRIS was made available through a dedicated website to be accessed by different users. The problems and prospects of the HRIS are enumerated in this article.

Reference

Sharma, S.D., *et al.,* (2002). Requirement Analysis Document, Integrated National Agricultural Resources Information System, IASRI, New Delhi.

ANNEXURE I

Fruits

• Mango	• Grape	• Banana
• Guava	• Pomegranate	• Citrus
• Papaya	• Pineapple	• Sapota
• Annona	• Apple	• Lichi
• Ber	• Amol	•

Vegetables

• Potato	• Tomato	• Onion
• Chilli	• Brinjal	• Garlic
• Watermelon,	• Beans	• Capsicum
• Cabbage	• Cauliflower	• Cucumber
• Ridgegourd	• Bottlegourd	• Bittergourd
• Gherkin	• Mushroom	• Peas
• Okra	•	•

Flower Crops

• Rose	• Carnation	• Tuberose
• Chrysanthemum	• Aster	• Marigold
• Gladiolus	• Orchids	• Anthuriums
• Gerbera	• Crossandra	• Lilium
• Jasmine		

GEOINFORMATICS APPLICATIONS IN AGRICULTURE, 2007
Eds. : A.K. Singh and U.K. Chopra; pp. 121-137
New India Publishing Agency, New Delhi (India)
E-mail : newindiapublishingagency@gmail.com
Web: www.bookfactoryindia.com

5

Geoinformatics in Support of Food Security and Poverty Alleviation: A Perspective Based on National Wasteland Mapping Project

S K Srivastava, CBS Dutt, R Nagaraja
S Bandyopadhayay, H C Meena Rani
V S Hegde & V Jayaraman

Key words : Wasteland Reclamation, Poverty Indicators, Food Insecurity, Alternate Livelihood Systems, Rural Infrastructure Index.

Introduction

The countries of Asia and the Pacific region accounts for nearly two-thirds of the chronically poor and undernourished in the world. FAO estimates indicate that, by 2010, Asia will still account for about one-half of the world's malnourished population (FAO 2000). In Asia, poverty has mainly been rural phenomenon and nearly three-fourth of the poor live in rural areas, with large majority of them dependent on natural resources for employment and income. South Asia, which had a poverty incidence of 43 percent (or about 520 million people), contributed about 40 percent of the world's poor. Development of natural resources thus offers a potentially enormous means of poverty reduction.

In developing countries poverty and environmental deterioration are often viable in proximity to each other, and have led many to infer that a two-way causality exists between human and environmental degradation. Environmental degradation usually occurs when production and consumption activities of growing population irreversibly weaken nature's recycling capabilities. These economic activities are also attributed to the development of markets, the advent of modern technology, and the spatial integration of inaccessible areas to market systems. In many developing countries, however, environmental degradation such as soil erosion, deforestation and pollution are most visible around poor settlements, leading some policy makers to highlight the direct links between poverty and the environment.

One has to recognize that both poverty and the environment are descriptions of states of human and natural resource attributes and cannot be reduced to simple one-dimensional cause-effect relationships. Apart from conceptual difficulties in modeling linkages, another handicap has been the absence of adequate or reliable data sets on poverty and environment characteristics. The challenge for operational research is exploring how circumstantial evidence and inductive logic can be used to explain the nature of interactions between the two states. Based on nationwide wasteland statistics vis-à-vis state and district wise poverty estimates available with the concerned agencies, this study is one such attempt, and evaluates the incidence of poverty with the backdrop of natural resources degradation and role of institutional interventions with illustrations from India.

Nationwide Wasteland Mapping

In the recent years, there has been growing needs by planners, to know spatially the regions with the co-existence of poverty and environmental degradation, their precise distributions and the impact of various poverty alleviation measures *vis-à-vis* environmental management.

It is in this context, the importance of satellite remote sensing & Geographical Information System (GIS) in providing such information is well documented (Srivastava *et al.,* 2002).

In India, for targeting environmentally degraded lands to initiate poverty alleviation programme, at the behest of Ministry of Rural Development, Government of India, National Remote Sensing Agency (NRSA), Department of Space carried out the nation-wide wasteland mapping which contain details on different type of wastelands, down the line at the district and even to village levels. Using GIS technique, village boundaries are integrated with micro-watershed boundaries and wasteland maps. The detailed methodology of nationwide wasteland mapping is discussed in Wasteland Atlas 2002 (Wasteland Maps of India, Ministry of Rural Development, Government of India, 2000).

Presently, these maps, serving as primary input in planning reclamation measures, micro-level inventory and monitoring of the wastelands reclamation measures, are being used by Department of Land Resources of Ministry of Rural Development, State Forest and Agriculture Departments, District Rural Development Agencies (DRDAs), NGOs etc for various institutional interventions aimed at poverty alleviation. For example, wasteland maps provide valuable inputs to the Integrated Wastelands Development Project (IWDP), which has been under implementation since 1989-90. The scheme is being implemented on watershed basis & is focused on employment generation in rural areas besides enhancing people's participation in the wasteland development programme. Similarly, there are other rural development programmes: Drought Prone Areas Programme (DPAP), Desert Development Programme (DDP), besides formulating land reforms policies, which will be supported by village level wastelands information. With the current emphasis on creating digital environment to enable e-governance especially in a sector like rural poverty alleviation, digital archives of village level wastelands is of great significance to harness the benefits of ICTs at grassroots levels.

Incidence of Poverty, Food Insecurity and Wastelands-Need for Disaggregated Poverty Mapping

In the simplest thinking, one could argue, ceteris paribus, that poor individuals, in their search for food, would convert forests to farms, grow food on steep slopes and degrade marginal farmlands. The derivative of Wasteland map generated by NRSA, areas showing marginal agricultural lands which are comparable to spatially broad level incidence of poverty map at State level, has also captured this (Figure 1).

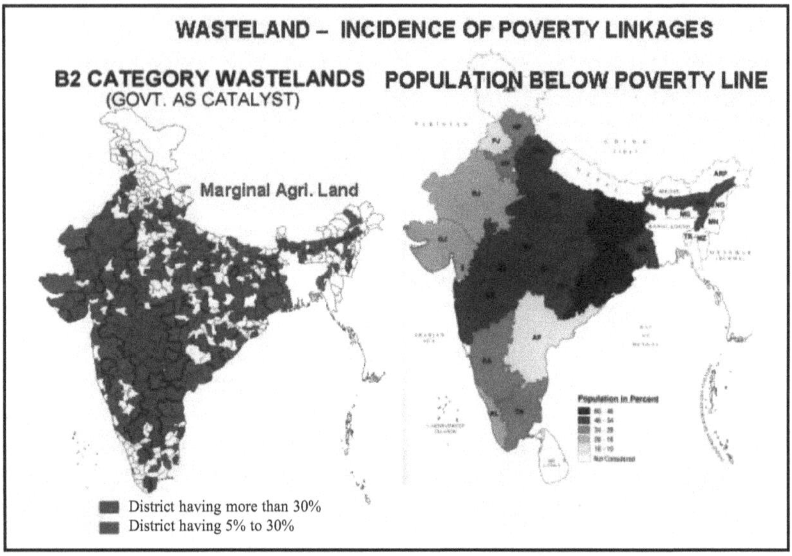

Figure 1. Marginal lands (*Source:* NRSA) and incidence of poverty (*Source:* Food Insecurity Atlas of Rural India)

In poor economies the pressure of population would force agriculture on to more and more marginal lands, permanently impairing nature's regeneration capabilities, this argument would lead one to conclude. Finite natural resources would thus face increasing problems of negative externalities, and ecological systems would be irreversibly degraded in their physical capabilities. To examine this simplistic assumption, wasteland maps provide valuable insights. Wastelands at state levels are listed along with other poverty and other socio-economic indicators in Table 1.

Despite more than 70 per cent population in the rural India dependent on natural resources, the relationship between wasteland and poverty is seen to be very complex (Figure 1).

In fact, at state level, the occurrence of wastelands does not seem to be connected with incidence of poverty.

- Bihar, for example, is characterized by just 6 % wastelands with 57% population below poverty line.
- Jharkhand, with 19% wastelands, has got more than 60% population below poverty line.
- Uttar Pradesh has 9% wastelands with more than 44% population below poverty line.
- Similarly, West Bengal with just over 6 % wastelands has got more than 40 % population below poverty line.

There is another set of states wherein the incidences of poverty as well as wastelands both are equally high.

- For example, Assam with more than 25 % wastelands has got more than 45 % population below poverty line.
- Similarly, Rajasthan is characterized by more than 30 % wastelands and 26 % incidence of poverty.

On the other hand, there are States like Punjab with just 4% wastelands and 11% population below poverty line, followed by Andhra Pradesh, Haryana, Kerala, Gujarat etc.

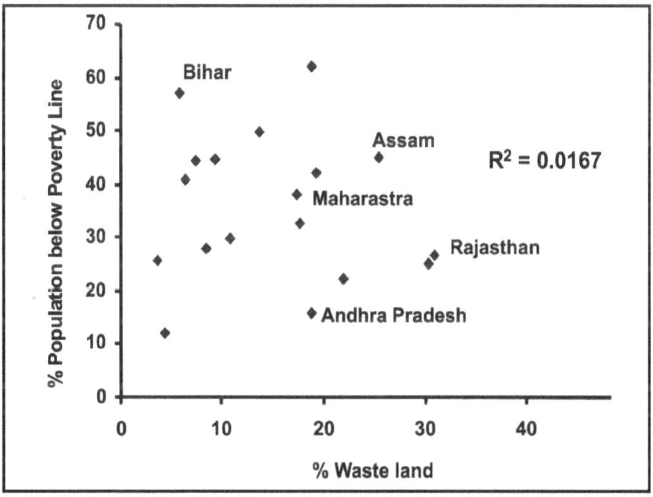

Figure 2. Wastelands & Incidence of Poverty linkages at State level

At district level, the relationship between incidence of poverty and wastelands has increased significantly for almost all the States. In case of Andhra Pradesh, for example, it is depicted in Figure 2 and Table 2. It is important to note that regression coefficient (R^2) between incidence of poverty and wastelands increased from 0.0167 at State levels to as good as around 0.5 for quite a few States at the district level. It is expected that regression coefficient may increase further if it is derived at village level. It brings out clearly that a meaningful relationship between incidence of poverty and wastelands exists at much higher level, may be at village level. The direct links between poverty and resource degradation requires further ground-level research and thus calling for dis-aggregated wasteland mapping to establish closer linkages with incidence of poverty.

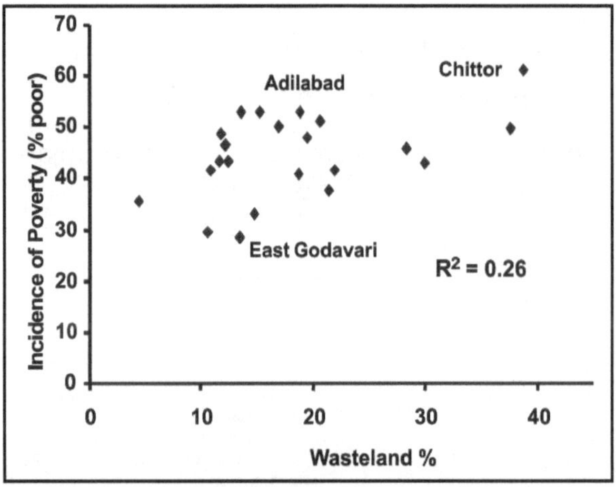

Figure 3. Wastelands & Incidence of Poverty linkages at District level - Andhra Pradesh

Role of Public Policy and Institutional Interventions

The linkages between poverty and environmental degradation are, however, not just governed by the physical limits of ecosystems, but rather, by the income strategies of the poor. Driven by public policies and institutional interventions, economic and spatial integration of markets occur and several new marginal income-earning opportunities become available in the informal sector of the economy. Consequently, the dependence of the poor on the natural resource base for livelihoods may actually get reduced. It is in this context that linkage between poverty and renewable natural resource degradation got altered. In view of this, it is required to evaluate the role of economic and institutional policies in altering labor and capital flows between and within regions (Figure 4).

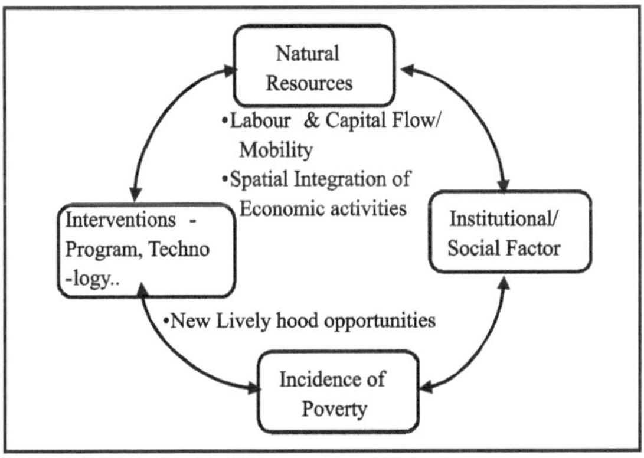

Figure 4. Poverty, Natural Resources, Institutional Interventions and Societal Dynamics - Linkages

Yet another aspect is, production and exchange through impersonal markets is strongly preferred to subsistence activities by poor producers and consumers alike. Although family decisions and mutual responsibilities may continue defining some aspects of household economic behavior, market exchange irreversibly supplants reciprocity and redistribution as the means of acquiring income and wealth (Dalton 1981). Under these circumstances, except for a small group of the ultra poor, most individuals and groups could be expected to strive for income security rather than just food security; cash being necessary to purchase non-traditional food and consumption items that reflect changing utility functions. A desire for income security has lead to creative income earning efforts through a myriad of highly divisible, labor intensive, low skill vocations involving wage labor and self-employment in rural India (Jagannathan 1987, Pean 1989). All these altered the relationship between poverty and natural resources degradation.

The availability of these (often unrecorded) income-earning opportunities reduces the necessity of the poor to mine land and water resources close to their homes for survival. Equally important is whether the poor have the power to gain access or to use available natural resources. Poverty is very often visible as unequal access to land and capital because asset ownership is inequitable (Table 1). Moreover, monetization also appreciates land and water values to society, and what were common property rights begin to get privatized or retain de facto access to productive and water resources because of social and institutional factors, such as changing tenurial relations, laws and regulations, social and political relations both within and outside villages. Jodha (1986), for example, has documented how common values, leaving the poor de facto access to only marginally productive wastelands.

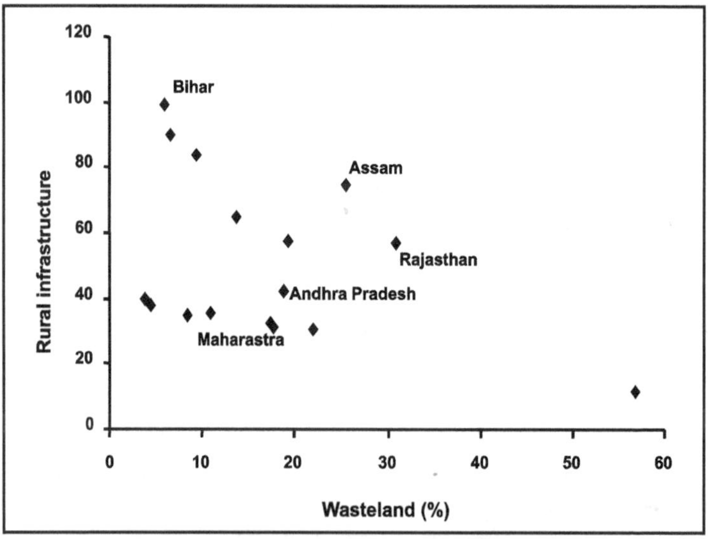

Figure 5. Rural infrastructure index and wasteland

Apart from market incentive structures and institutional constraints, India has attempted to improve the welfare of the rural poor through social and economic policies. Before drawing any policy conclusions on poverty-environment linkages, one has to look beyond just the immediate physical environment, and correlate the largely individualistic survival strategies pursued by the poor to generate income and food security with:

- Public policies designed to provide basic needs services to poor citizens, because of which economic expectations of the poor begin improving;
- Natural resource utilization by poor communities, taking into account spatial choices that are available for earning livelihood between farming, off-farm activities and non-farm activities.

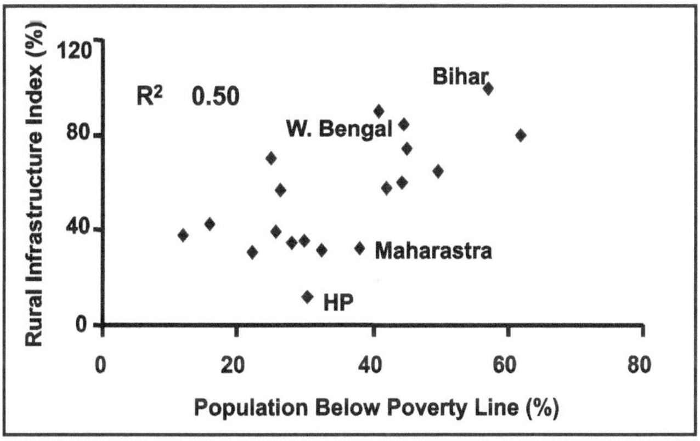

Figure 6. Rural infrastructure index vs. incidence of poverty

The issue for research becomes disentangling complex casual processes, and explaining whether poverty, public policies, institutions or combinations of them cause changes in natural resource use.

Spatial Integration of Rural Poor to Market and Creation of Alternate Livelihood Systems

Natural Resources data on soils, topography, vegetation, water etc, have been often utilized to assess the economic potentials of land areas. It is however equally important to understand how macro-economic variables could determine the dynamics of poverty and natural resources degradation relationship. In this context, it is important to examine the role of certain factors especially the education and rural infrastructure index (Appendix I), which enable the poor to integrate them with market and other off farm opportunities. Infra-structural investments, notably road construction, have played major roles in changing natural resource use. Roads have increased areas under commercial agriculture, enabled entrepreneurs to invest in natural resources development, and generally increased the mobility of capital and labor.

- In states like Bihar, West Bengal, Uttar Pradesh and Orissa, while rural infrastructure has been very poor (Table 1), the level of education (except for West Bengal) has also been equally poor. Lack of these factors constrains the spatial integration of poor to the market and other opportunities.
- The states like Himanchal Pradesh, Gujarat, Tamilnadu and Maharastra, by virtue of having higher rural infrastructure though the level of education is still not high, have succeeded integrating the poor to the market and exposed them to non-farm opportunities. The rural hinterland in these states appears to have achieved spatial integration, creating large markets for commodities and labor.
- The integration of poor to market and non-farm opportunities in Kerala is driven more by literacy followed by rural infrastructure and deficit in food production.

The spatial dynamics of poverty seem to be influenced by the rural infrastructure, availability of transport links, and the growth of production and consumption linkages (Figure 5 and 6). The economic stimuli however have also been influenced by several other factors: the structure of product and factory markets, policies that encouraged private investments in natural resources development, and the size and direction of public investments.

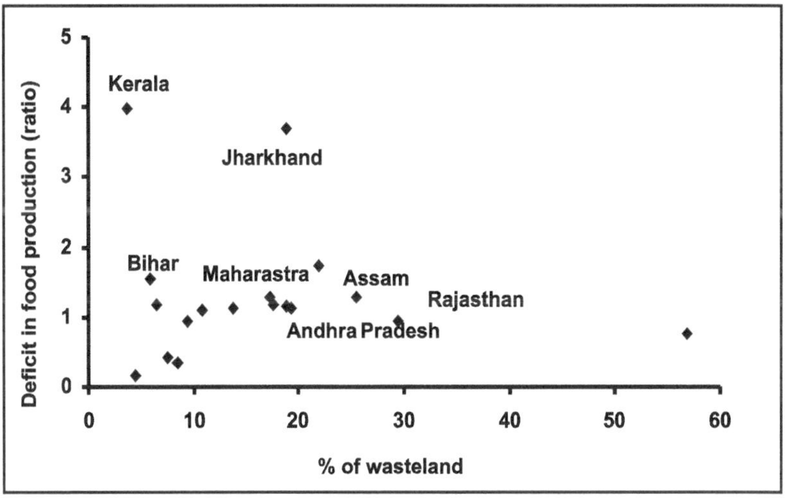

Figure 7. Deficit in food production represented by the ratio of consumption to production of cereals (1993 -94) vs. wastelands

Food Insecurity, Wastelands and Scope for Interventions

Poverty alleviation and food security are truly inseparable. Food security, providing the economic access to the basic nutritional food, is a means of poverty alleviation. In case of the total dependence of the rural population on natural resources for food and livelihood, wastelands have direct relevance to food security. The figure 7 does not substantiate that the deficit in food production (Food Insecurity Atlas of Rural India, WFP & MSSRF, 2001) is directly connected to wastelands (Table 1).

- Jharkhand, Assam and Rajasthan with very high degree of food insecurity as well as large incidence of wasteland, provide unique case for institutional interventions wherein agricultural expansion/intensification could form the basis of poverty alleviation,
- On contrary, Kerala and Bihar, both being also highly food insecure States need to have interventions in other areas for poverty alleviation, as wastelands are quite low in these States,

Table 1. Wasteland statistics and indicators of poverty and food insecurity - India

State	Deficit in food production	% Waste land	% Below Poverty Line	Rural infrastructure index (%)	Rural literacy (%) 2001
AP	1.16	18.81	15.92	42.3	55.33
Assam	1.29	25.52	45.01	74.6	60.92
Bihar	1.55	5.9	56.93	99.2	44.42
Chattisgarh	0.42	7.53	44.38	60.0	76.23
Gujrat	1.74	21.95	22.18	30.8	58.53
Haryana	0.33	8.45	28.02	34.9	64.0
HP	0.76	56.87	30.34	11.8	74.38
Karnataka	1.11	10.87	29.88	35.8	60
Kerala	3.99	3.73	25.76	39.7	90.05
Jharkhand	3.7	18.89	62.0	80.0	46.26
MP	1.12	19.31	42.05	57.4	58.1
Maharastra	1.27	17.38	37.93	32.4	71
Orissa	1.13	13.71	49.72	64.6	66.44
Punjub	0.16	4.42	11.95	37.7	65.0
Rajasthan	1.25	30.87	26.46	56.9	56.0
Tamil Nadu	1.18	17.7	32.48	31.4	67.0
UP	0.94	9.4	44.54	84.1	54.0
Uttaranchal	0.85	30.27	24.98	70.0	61.0
WB	1.18	6.44	40.8	89.9	64.0

- The interventions in food secured States Punjab, Chattisgarh, Himachal, Haryana, Uttaranchal, Uttar Pradesh and Karnataka need to be focused on off-farm income generation.

It is therefore important to highlight that the economic policies and institutional interventions need to focus on spatial integration of rural poor to the market forces and creation of alternate livelihood systems. The strategy however could be highly disaggregated depending on various *in situ* conditions including the relationship between poverty and natural resources degradation.

Conclusions

For many developing countries like India, natural resource endowments are the most significant component of national wealth. There is a danger that poorly designed policies, rapidly increasing populations and poverty could irretrievably damage their capital stock. Data on both socio-economic and natural resource variables are usually unavailable at sufficient detailed level (cadastral information or even individual settlements) to examine this. Collecting these data sets are also unaffordable because of competing demands for limited budgetary resources. What this desk study tried to show was that despite these practical problems, existing sources of wasteland statistics in conjunction with other socio-economic data could lead to a more aggregative, regional-level analysis of natural resource use-poverty-public policy linkages and suggest the appropriate institutional interventions for poverty alleviation.

In conclusion, many macro-economic policy instruments, such as investments in rural infrastructure, education etc result in stock changes of physical capital in identifiable areas and thus possibly could alter the relationship between incidence of poverty and natural resources degradation. The economic policy instruments, if translated from conventional monetary aggregates to spatial 'stocks' or areal densities (Beckmann and Pun 1985, Jagannathan Vijay N, 1989), it is possible to evolve the appropriate institutional interventions, which could lead to substantial poverty reduction as well as enrichment of natural resources base.

APPENDIX - I
Rural Infrastructure Index

Rural infrastructure consists of certain key indicators, which facilitate livelihood access. They reflect certain basic amenities available in rural areas such as:

- Road length per thousand persons
- Percentage of villages without electricity
- Percentage of households without electricity and
- Percentage of households without piped water

Road length facilitates transport of more food to the villages. Electrification helps better production. Electrification also helps processing industries and other non-agricultural enterprises, which would enhance livelihood access and food access.

Better water supply and availability of electricity in the households would enhance the capacity for enterprise. Productivity of the rural population would go up if there were a reduction in the time and effort involved in carrying water and fuel from long distances. Rural infrastructure has a direct bearing on livelihood access and food access. For livelihood access, markets and credit institutions are also very important.

The chosen indicators are first converted into indices and then averaged together to get the composite index. The method of calculating the index is as follows for all the indicators, except 'Road length per thousand persons'

$$1_{ij} = (X_{ij} - X_{imn}) / (X_{imx} - X_{imn})$$

where

X_{ij} = i^{th} Rural infrastructure indicator in the jth state

X_{imx} = i^{th} Rural infrastructure indicator with maximum value among all the states indicating the worst situation (This gets the value of one)

Table 2. Wasteland statistics and indicators of poverty and marginalization - Andhra Pradesh, India

District	Incidence of Poverty (poor %)	Wasteland (%)	Small & Marginal holding (%)
Adilabad	53.1	15.23	54
Ananthpur	50	16.9	63.7
Cuddapah	43.1	29.93	84.6
East Godavari	28.6	13.41	55.4
Guntur	33	14.72	87.2
Karimnagar	43.2	12.43	84.2
Kammam	41.7	10.9	68.3
Krishna	29.4	10.56	90.1
Kurnool	41.7	21.97	67
Mahaboobnagar	52.8	13.54	67.2
Medak	48.8	11.73	74.4
Nalgonda	46.6	12.14	74
Nellore	49.7	37.61	97.8
Nizamabad	41.0	18.7	82.4
Prakasham	37.8	21.46	87.8
Ranga Reddy	47.8	19.5	67.9
Srikakulam	53.1	18.89	92.2
Vishakhapatnam	45.9	28.24	74.7
Vizianagaram	51.3	20.63	85.8
Warangal	43.2	11.63	80.8
West Godavari	35.7	4.38	71

X_{imn} = i^{th} Rural infrastructure indicator with minimum value among all the States indicating the best situation (This gets the index value of Zero),

In respect of Road length per thousand population, the following formula was used to get the Index of road length.

$I_{ij} = (X_{imx} - X_{ij}) / (X_{imx} - X_{imn})$

I_{RI} = Index of Rural Infrastructure is calculated as follows:

$I_{RI} = \{ \lambda (I_{ij}) / n\} \times 100$

'i' = 1 to 4 Rural infrastructure indicators

'j' = 1 to 16 states in the country

The composite Index is the average of all the four indices. Each index measures the distance of the state from the worst possible situation, compared to the distance between the best and the worst states. A composite rural infrastructure index of 99 percent for Bihar means that Bihar has to travel 99 percent of the distance to reach the level of state with best infrastructure.

References

Food and Agricultural Organisation (FAO), (1996), Food for All, Report by FAO on the occasion of the World Food Summit, 13-17, November.

Food and Agricultural Organisation (FAO), (2000), The State of Food Insecurity in the World, Second edition.

Jodha, N.S (1995),'Common Property Resources and the Environmental Context: Role of Biophysical versus Social Stresses', *Economic and Political Weekly*. December 23, pp. 3278-3283.

Dalton, George, 1981, Research in Economic Anthropology, Vol 4, New York, Jai Press.

Jagannathan Vijay N, 1987, Informal Markets in Developing Countries, Oxford University Press, New York.

Jagannathan Vijay N, 1987, Planning new towns : the Durgapur experience, *Economic and Political Weekly*, March 28.

Pean, Leslie, 1989, Working paper on the Urban Informal sector in Sahel, Infrastructure Operations, Sahelian Department World Bank.

Jagannathan Vijay N, 1989, Poverty, Public Policies and the Environment, Environment Working Paper No. 24, the World Bank, Policy Planning and Research Staff, Washington, US.

Jodha, N.S., 1986, Common Property Resources and Rural Poor in Dry Regions of India, *Economic and Political Weekly*, Vol, XXI No.27, July 5, 1986.

World Bank, Rural Development World Bank Experience 1965-86 (1988), Washington, US.

Beckmann, M and T Pun, Spatial Economics: Density, Potential and Flood, Amsterdam, North Holland; studies in Regional Sciences and Urban Economics, 1985.

Food Insecurity Atlas of Rural India, 2001,World Food Progtamme and MS Swaminathan Research Foundation Centre for Sustainable Agriculture and Rural Development, Chennai, India.

Government of India, Ministry of Rural Development and National Remote Sensing Agency, 2000, Wasteland Atlas of India.

Srivastava, S K, Jayaraman, V and Parihar, JS, 2002, Strategy and Framework for Capacity Building in Remote Sensing and GIS Applications for Poverty Alleviation and Food Security in Asia & the Pacific, Technical Report, ISRO/EOS/2002,UN-ESCAP/ISRO Regional Seminar and Group Training on Monitoring and Assessment of Rice Crop in the ESCAP Region, Feb 12-15, 2002 Ahmedabad, India.

❑❑❑

GEOINFORMATICS APPLICATIONS IN AGRICULTURE, 2007
Eds. : A.K. Singh and U.K. Chopra; pp. 139-174
New India Publishing Agency, New Delhi (India)
E-mail : newindiapublishingagency@gmail.com
Web: www.bookfactoryindia.com

6

Geoinformatics for Monitoring Glaciers and Glacial Lakes

R.M. Bhagat, Chitra Sood, Vaibhav Kalia, Pradeep Mool, S. Bajracharya & Basanta Shrestha

Key words : Glaciers, Glacial Lakes, Glacial Lake Outburst Floods (GLOF), Himachal Pradesh, Glacial Inventory, Glacier Recession.

Introduction

Glaciers are large masses of snow, re-crystallized ice and rock debris that accumulate in great quantities and begin to flow outwards and downwards under the pressure of their own weight. Glaciers form when yearly snowfall in a region far exceeds the amount of snow and ice that melts in a given summer. In this way, massive quantities of glacial material accumulate in relatively small periods of geologic time. The glaciers exist in environments of high and low precipitation and in many temperature regimes; they are found on all the continents except Australia and they span the globe from high altitudes in equatorial regions to the polar ice caps. Presently, 10% of the Earth's land area (5.8 million square miles) is covered with glaciers storing about 75% of the world's freshwater. However, during the last Ice Age, glaciers covered 32% of the total land area of the planet.

Glaciers play an important role in maintaining ecosystem stability as they act as buffers and regulate the runoff water supply from high mountains to the plains during both dry and wet spells. The Himalayan glaciers account for about 70% of the world's non-polar glaciers and affect the lives of millions of people in several countries: China, India, Pakistan, Afghanistan, Nepal, Bhutan and Bangladesh. Their runoff feeds two of the oldest rivers in the world, the Indus and the Ganges, whose tributaries carry precious water for 500 million people on the northern Indian plains. Most of the glaciers in the Himalayas are of a summer-accumulation type that is major accumulation and ablation take place simultaneously during summer (Fujita *et al.*, 1997). It has been estimated that an area of about 32,000 sq. km is under permanent cover of ice and snow in the Himalayas (Negi, 1991). This amounts to about 17% of the total geographical area of the Himalayas. Higher concentration of glaciers in the Himalayas lie in the regions with the higher mountain peaks.

In the past few decades, global climate change has had a significant impact on the high mountain environment: snow, glaciers and permafrost are especially sensitive to changes in atmospheric conditions because of their proximity to melting conditions. In fact, changes in ice occurrences and corresponding impacts on high-mountain systems could be among the most directly visible signals of global warming. This is also one of the primary reasons why glacier observations have been used for climate system monitoring for many years (Haeberli 1990; Wood 1990). It has been estimated that sixty-seven percent of glaciers are retreating at a startling rate in the Himalayas and the major causal factor has been identified as climate change (Mayewski and Jeschke, 1979; Ageta and Kadota, 1992; Yamada, 1996). Glacial melt will affect freshwater flows with dramatic adverse effects on biodiversity, people and their livelihoods, with a possible long-term implication on regional food security. It is posited that the accelerated melting of glaciers will cause an increase in river levels over the next few decades, initially leading to higher incidence of flooding and land-slides (IPCC, 2001). But, in the longer-term, as the volume of ice available for melting diminishes, a reduction in glacial runoff and river flows can be expected (IPCC 1996, Wanchang *et al.*, 2000). In the Ganga, the loss of glacier melt water would reduce July-September flows by two thirds, causing water shortages for 500 million people and 37 percent of India's irrigated land (Jain 2001; Singh *et al.,*1994).

Further, the problems of water stress is already prevalent in the region due to increasing demands of domestic, agriculture, industry and a growing population. Any reduction in the availability of freshwater could have serious consequences for the economy, the environment and the daily lives of many millions of people within the affected basins and beyond. Rapid urbanization, population explosion and haphazard development are the main cause for the increasing pressure on our vulnerable fresh water resources. The demand for these limited water resources is rapidly increasing for agricultural, industrial, domestic and environmental uses. Thus, any reduction in the availability of freshwater could have serious consequences in matter of food security, people's livelihoods, industrial growth and environmental sustainability of the region. A reduction in water from the mountains would further affect the economy of the region by limiting the energy from hydropower plants and hampering industrial productivity (Johannesson, 1997). The prediction that "glaciers in the region will vanish within 40 years as a result of global warming" and that the flow of Himalayan rivers will "eventually diminish, resulting in widespread water shortages" (*New Scientist,* 1999, 2003) is equally disturbing.

In general, the area in Himalayas above an elevation of 4,000 masl is mostly covered by snow and ice throughout the year. The glaciers, some of which consist of a huge amount of perpetual snow and ice, create many glacial lakes. Glacial lake outburst floods (GLOFs) are catastrophic discharges of water resulting primarily from melting glaciers. An accelerated retreat of the glaciers in recent times has led to an enlargement of several glacial lakes. The GLOF is the most devastating flood-producing phenomenon, which occurs in the Himalayan region from time to time. In the last half-century, several glacial lakes have developed in the Hindu Kush-Himalayas and Tibetan Himalayas. The glacial lakes are formed on the glacier terminus due to the recent retreating processes of the glaciers. The majority of these glacial lakes are dammed by unstable moraines, which were formed by the glaciations of the Little Ice Age. Occasionally, the lake happens to burst and suddenly releases an enormous amount of its stored water, which causes serious floods downstream along the river channel.

GLOF refers to the sudden outburst of lakes, dammed by glacier ice or moraines, producing flows of water that are often of an order of magnitude greater than normal rain-derived peak flows and may travel tens of kilometres downstream, transporting colossal amounts of debris. The occurrence of GLOF in high mountain areas poses many problems for inhabitants and their infrastructure. First, the loss of human life is extremely high, second, agricultural crops and land are washed away on a large scale, third, hydroelectric projects and roads are totally destroyed or damaged, then rock-fill dams may be breached if the inflow of the torrent of debris cannot be accommodated over the spillway, and lastly large reservoirs are rapidly filled with sediments. Several incidents of GLOF causing national economic losses worth millions have been reported from various parts of Hindu Kush Himalayas (Dwivedi, 2000; Mool *et al.,* 2001).

The bursting of moraine-dammed lakes is often due to the breaching of the dam by the erosion of the dam material as a result of overtopping by surging water or piping of dam material. Earthquakes leading to the slumping of dam material may also cause the bursting of the lake. The drainage of ice-dammed lakes may be due to: flotation of the ice dam, pressure deformation, melting of tunnels through or under the ice, and drainage associated with tectonic activity.

Glaciers and glacial lakes are generally located in remote areas, where access is through tough and difficult terrain and takes several days to reach the area. The inventory and field investigation of glaciers, glacial lakes and glacial lake outburst floods using conventional methods, requires extensive time and resources together with hardship in the field. Inventories and monitoring of the glaciers, glacial lakes and extent of GLOF impact along the downstream can be done quickly and precisely using time series satellite images and aerial photographs. Use of these images and photographs for the evaluation of physical conditions of the area provides greater accuracy. The multi -stage approach using remotely sensed data and field investigation increases the ability and accuracy of the work. A digital database of glaciers and glacial lakes therefore, is necessary to identify the potentially dangerous glacial lakes. Geographic Information System (GIS) is the most appropriate tool for spatial data input and attribute data handling. It is a computer-based system that provides the capabilities to handle data input, data management (data storage and retrieval), data manipulation and analysis, and data output. Visual and digital image analysis integrated with techniques of GIS are very useful for the study of glaciers, glacial lakes and GLOFs.

In view of this fact, therefore, monitoring of glaciers is not only crucial to assess the future availability of water sources as they provide sufficient water in melt-off, but also to understand the phenomenon of glacial lake outburst floods (GLOF) to develop an early warning system for GLOF events for avoiding flash flood tragedies common in the mountain regions. Besides, it can provide an estimation of the water reserves for agricultural production and evaluating the impact of climate change/global warming on mountain agriculture. Regular monitoring of glaciers and glacial lakes is useful in understanding the physical and climatological characteristics of the region and serve as a useful reference for future research.

Methodology

For inventorying glaciers and glacial lakes, a digital database of glaciers and glacial lakes using geographic information systems (GIS) is developed. A digital database is necessary for the monitoring of glaciers and glacial lakes and to identify the potentially dangerous lakes. GIS is the most appropriate tool for spatial data input and attribute data handling. It is a computer-based system that provides the following four sets of capabilities to handle geo-referenced data: data input, data management (data storage and retrieval), data manipulation and analysis (Arnoff, 1989).

The basic materials required for the compilation of an inventory of glaciers and glacial lakes are remote sensing data, large-scale topographic maps and aerial photographs. The combination of digital satellite data and the digital elevation model (DEM) of the area is also used for better and more accurate results for the inventory of glaciers and glacial lakes.

Use of Remote Sensing and Geographic Information System

Creating inventories and monitoring of the glaciers, glacial lakes, and extent of GLOF impact downstream can be done quickly and correctly using satellite images and aerial photographs. Use of these images and photographs for the evaluation of physical conditions of the area provides greater accuracy. The multi-stage approach using remotely sensed data and field investigation increases the ability and accuracy of the work.

Remote sensing technology makes use of the wide range of the electro-magnetic spectrum (EMS). Most of the commercially available remote-sensing data are acquired in the visible, infrared, and microwave wavelength portion of the EMS. There are different types of commercial satellite data available. Digital data sets of the Land Observation Satellite (LANDSAT) Thematic Mapper (TM), IKONOS, QUICKBIRD etc; in India - Indian Remote Sensing Satellite Series (IRS) Linear Imaging and Self Scanning Sensor (LISS) are widely available for studying glaciers. The data sets of Système Probatoire Pour l'Observation de la Terre (SPOT) Multi-Spectral (XS) and SPOT Panchromatic (PAN) can also be used depending upon the availability and money involved. Different techniques of digital image enhancement and spectral classification of ground features can be exploited for the study of glaciers, glacial lakes and further identification of the potentially dangerous lakes.

Snow and ice are the frozen states of water. The satellite remote sensing systems' data are available in more spectral bands (e.g. LANDSAT TM in seven bands). It is now possible to differentiate snow and cloud easily in the middle infrared portion of the spectrum, particularly in the 1.55 - 1.75 ìm and 2.10 - 2.35 ìm wavelength bands (bands 5 and 7 of LANDSAT TM). As shown in Fig. 1, in these wavelengths, the clouds have a very high reflectance and appear white on the image, while the snow has a very low reflectance and appears black on the image. In the visible, near infrared, and thermal infrared bands, spectral discrimination between snow and clouds is not possible, while in the middle infrared it is. The reflectance of snow is generally very high in the visible portions and decreases throughout the reflective infrared portions of the spectrum. The reflectance of old snow and ice is always lower than that of fresh snow and clean/fresh glacier in all the visible and reflective infrared portions of the spectrum. Compared to clean glacier and snow (fresh as well as old), debris covered glacier and very old/dirty snow have much lower reflectance in the visible portions of the spectrum and higher in the middle infrared portions of spectrum.

To identify the physical conditions of individual glaciers and glacial lakes, different image enhancement techniques are useful, however, complemented by the visual interpretation method (visual pattern recognition) and with the knowledge and experience of the terrain conditions. Different colour composites (FCC) and individual spectral bands images highlight different land-cover features using the knowledge of image interpretation keys: colour, tone, texture, pattern, association, shape, shadow etc.

The technique of integrating remote-sensing data with GIS does help a lot with identification and monitoring of lakes and glaciers. The DEM of an area generated, either using stereo satellite images, aerial photographs, or digitization of topographic map data, can play a big role in deciding the rules for discrimination of features and land-cover types in GIS techniques and for better perspective viewing and presentations. DEM itself can be used to create various data sets of the area (e.g. slope, aspect). For example, even though glacial lakes are covered by snow, the lake surfaces are flat, and glaciers, snow, and ice give some slope angle. In this case,

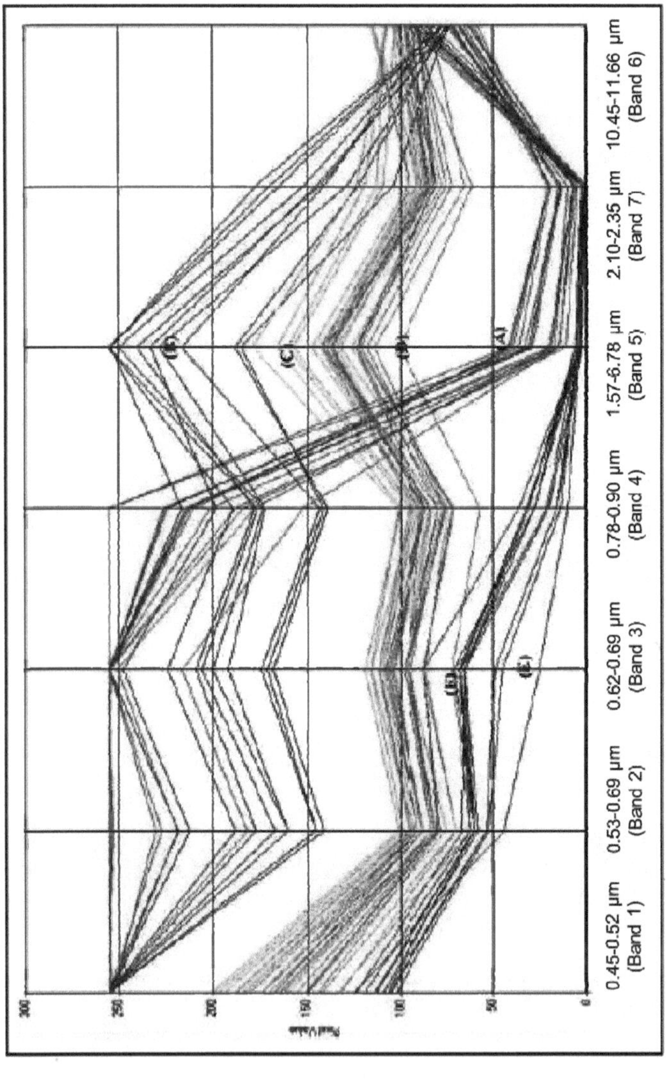

Figure 1. Spectral reflectance characteristics in terms of pixel value based on a September 22, 1992, Landsat TM seven-band data set. Red lines—clean glaciers and fresh snow (A), Black lines—clouds (B), Green lines—recent debris from GLOFs (C), Maroon lines—debris covered glacier (D), Blue lines—clean/melted (E) silty and/or partly frozen water (lake) (F)

decision rules for integrated analysis in GIS can be assigned, that is, if the slope is not so pronounced, then those areas are recognized as the frozen glacial lakes. DEM should be compatible and of reliable quality when compared with other data sets. The satellite images or orthophotos can be draped over the DEM for interpretation or presentation.

Inventory of Glaciers

The methodology for the mapping and inventory of the glaciers is generally based on instructions for compilation and assemblage of data for the World Glacier Inventory (WGI), developed by the Temporary Technical Secretary (TTS) at the Swiss Federal Institute of Technology, Zurich (Muller et al., 1977) and the methodology for the inventory of glacial lakes is based on that developed by the Lanzhou Institute of Glaciology and Geocryology, the Water and Energy Commission Secretariat, and the Nepal Electricity Authority (LIGG/WECS/NEA 1988). The inventory of glaciers and glacial lakes can be systematically carried out for the drainage basins on the basis of topographic maps and satellite images.

The glacier margins on each map are delineated and digitized on the satellite image and the identification, classification, and determination of stages of glaciers can be accomplished by referring topographic maps of the glaciated regions. The different attribute parameters required to be evaluated for making an inventory are: proper numbering of the glaciers, registration of snow and ice masses, and careful demarcation of the snow line etc. with possibility of errors minimized by using Accuracy Rating Table proposed by Muller et al., (1977). In the glacier inventory, different methods or a combination of methods are usually chosen for comparison with aerial photographs in order to assess the elevation of the firn line for different forms of glacier.

All the polygons representing glaciers and glacial lakes are numbered and points showing the location of glaciers and glacial lakes can be digitized. Attributes such as area, location (latitude, longitude) etc. can be derived from the spatial database. If other necessary digital spatial data layers, such as digital elevation models (DEM), are available, it is possible to generate terrain parameters such as elevation, slope, length etc as measuring units for glaciers and glacial lakes. Other attributes, such as aspect, mean length, elevation, map code, name, etc. were manually entered in the attribute database. Additional attributes, such as mean elevation, volume etc were derived using logical calculations.

The aerial extension of the glaciers is found with the help of geographic information systems (GIS). The delineated glacier area is digitized in the integrated land and water information systems' (ILWIS) format and the database is used to calculate the total area, length, mean width, orientation, elevation etc. To estimate the ice reserves, it is an utmost necessity to have the mean thickness of the glaciers. Since the mean glacier thickness data are not available, this is estimated from the equation developed for the Tianshan Mountains (Chaohai Liu and Liangfu 1986).

$$H = -11.32 + 53.21F^{0.3}$$

Where H = mean ice thickness (m) and F = area of glacier (km^2)

The ice reserves were estimated by multiplying the mean thickness by the area of the glacier. The categorisation of glaciers into different types viz., Mountain or valley (simple or compound basin) considering the area and depth of the glaciers (Muller et al., 1977).

Inventory of Glacial Lakes

Changes in climatic conditions have had an impact on the high mountain glacial environment. Many of the big glaciers have melted rapidly and given birth to a large number of glacial lakes. Due to the rapid rate of ice and snow melt, possibly caused by global warming, the accumulation of water in these lakes has been increasing rapidly. The isolated lakes above 3,500 masl are assumed to be remnants of the glacial lakes left due to the retreat of the glaciers. The glacial lake inventory on the analogy of glaciers can be systematically compiled for the drainage basins on the basis of topographic maps and satellite images.

The physical attribute parameters like numbering, longitude and latitude, area, length, width, depth, orientation, drainage condition (open or closed) and altitude were derived from the topographic maps, aerial photographs, satellite images, reports, field data, etc.

Attributes such as area, location (latitude, longitude) etc can be derived from the spatial database like glacier inventory.

Further classification of lakes into different types, erosion, blocked, moraine dammed, cirques is possible by visual interpretation techniques of satellite imagery as well as close monitoring of the lake. According to their stability, the glacial lakes are divided into three types: stable, potential danger, and outburst (when there have been previous lake bursts).

One important index for evaluating the stability of a glacial lake is its contact relation with the glacier. So an item of distance from the upper edge of the lake to the terminus of the glacier has been added and the code of the corresponding glacier registered. Since an end moraine-dammed lake is related to its originating glacier, this index is only referred to end moraine dammed lakes. As not enough field data exist, the average depth of glacial lakes is difficult to establish in most cases. Based on field data, and as an indication only, the average depth of a glacial lake formed by different causes can be roughly estimated as follows: cirque Lake, 10m; lateral moraine lake, 30m; trough valley lake, 25m; blocking lake and glacier erosion lake, 40m; lateral moraine lake, 20m. The water reserves of different types of glacial lakes can be obtained by multiplying their average depth by their area (LIGG/WECS/NEA 1988).

Potentially dangerous glacial lakes

On the basis of actively retreating glaciers and other criteria, the potentially dangerous glacial lakes were identified using the spatial and attribute database complemented by multi-temporal, remote-sensing data sets. Medium to large scale aerial photographs are also useful for detailed geomorphic studies and for evaluation of the active glaciers and potentially dangerous lakes.

Different triggering mechanisms of GLOF events depend on the nature of the damming materials, the position of the lake, the volume of the water, the nature and position of the associated mother glacier, physical and topographical conditions, and other physical conditions of the surroundings.

In general, based on geomorphic characteristics, glacial lakes can be grouped into three types: glacial erosion lakes, glacial Cirque lakes, and moraine-dammed lakes. The former two types of glacial lakes occupy the lowlands or emptying Cirques eroded by ancient glaciers. These glacial lakes are more or less located away from present-day glaciers and the downstream banks are usually made of bedrock or covered with a thinner layer of loose sediment. Both of these glacial lakes do not generally pose an outburst danger. On the other hand, the moraine-dammed glacial lakes have the potential for bursting.

Moraine-dammed glacial lakes, which are still in contact or very near to the glaciers, are usually dangerous. Peak discharge from breached moraine-damaged lakes just downstream from the moraine can be estimated from an empirical relationship developed by Costa (1985) is presented in Fig 2.

The criteria for identifying the potentially dangerous glacial lakes are based on field observations, processes and records of past events, geomorphic and geotechnical characteristics of the lake and surroundings, and other physical conditions like rise in lake water level, activity of supraglacial lakes, position of lakes, dam conditions, conditions of associated mother glacier, physical conditions of the surrounding area.

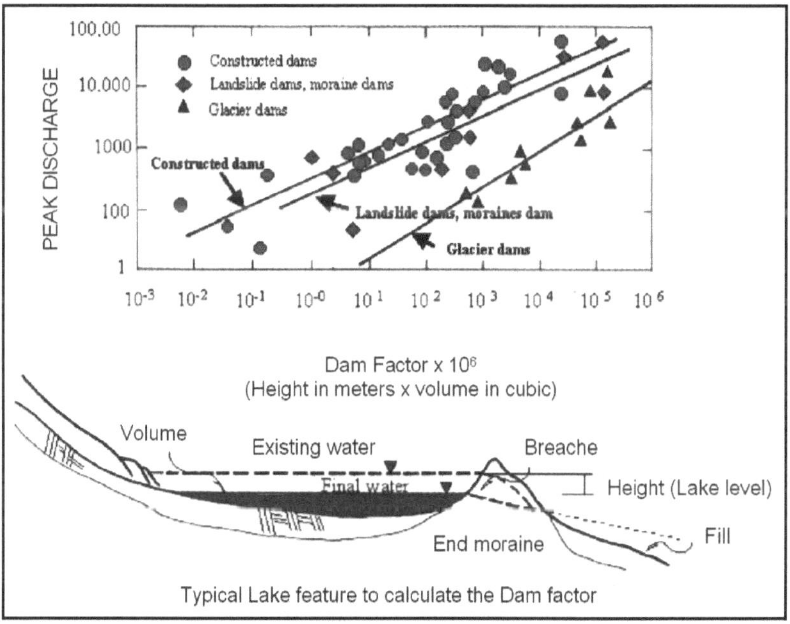

Figure 2. Mechanism of moraine dammed lake failure-Peak discharge from breached lake.

Case Study - Glaciers and Glacial Lakes in Himachal Pradesh

Himachal Pradesh is a mountain province in the Indian Himalayas covering an area of over 50 thousand sq. km, where mountains and hills occupy most of the land. It extends from the Shivalik hills in the south to the Great Himalayan range including a slice of Trans-Himalayas in the north. Geographically, the state of Himachal Pradesh is situated between 30°22′44″ and 33°12′40″ N latitude, and 75°45′55″ to 79°04′20″E longitude and occupies an area of 5.57 million ha. Himachal Pradesh is a hilly state with a general increase in elevation from west to east and south to north ranging from 350 m to 7000 m (Kayastha, 1971). The climatic conditions of the state vary from extremely hot to severe cold regions, a small area in north-western part of the state receives the highest precipitation as high as 2700 mm while dry temperate regions have a rainfall of around 124 mm. Administratively, the state is divided into 12 districts and 109 tehsils (Fig. 3).

As the name connotes (Him means snow), it is a land of snow-clad mountains, snow-fed rivers, icy lakes and springs. In general, about 6% of the area above 5,000 metres above mean sea level (masl) in Himachal Pradesh is mostly covered by snow and ice throughout the year forming the glaciers - the major storehouses of water. The four major rivers of the state: Beas, Ravi, Chenab and Satluj feeding millions of population downstream basically rely on the perennial supply of melt-water from the Himalayas (Sharma 2001).

In the past few decades, global climate change has had a significant impact on the high mountain environment; snow, glaciers and permafrost are especially sensitive to changes in atmospheric conditions because of their proximity to melting conditions. In the Indian mountain state of Himachal Pradesh, evidences of global warming could be clearly deciphered by

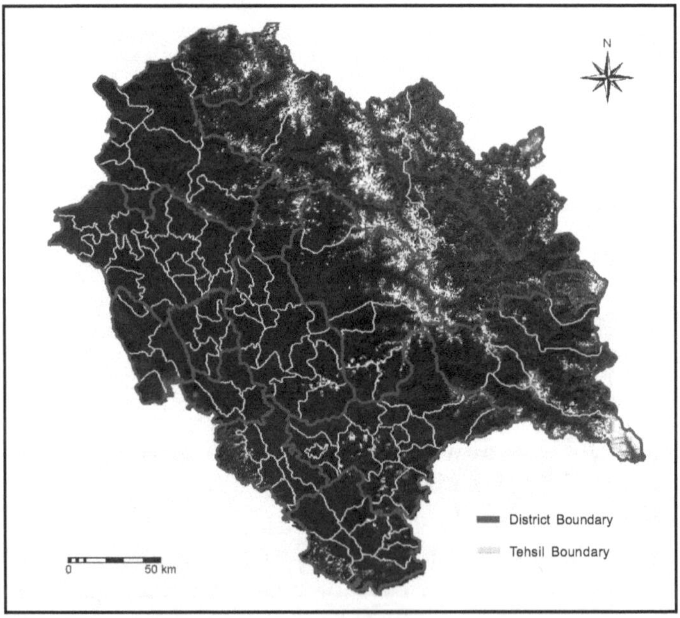

Figure 3. A map of Himachal Pradesh showing Boundaries of Districts and Tehsils.

changes like receding snowfall in the mountains, retreating glaciers and drying up of perennial hill springs, shifting of temperate fruit belt upward, adversely affected productivity of apples, shortening and shifting of Rabi season forward, disrupted rainfall pattern and more severe incidences of diseases and pests (Bhagat *et al.,* 2005).

The state is vulnerable to various hazards due to fragile geological conditions, great elevation differences, and steep sloping terrain. Apart from landslides and river erosion, the mountainous region is also quite susceptible to disastrous hazards due to Glacial Lake Outburst Floods (GLOFs). In the last half-century, several glacial lakes have developed in the Himachal Pradesh Himalayas and Tibetan Himalayas, which is a possible danger for lowland valleys of the State. This may be attributed to the effect of recent global warming.

Glaciers of Himachal Pradesh

A comprehensive inventory and a geo-database of glaciers and glacial lakes occurring in Himachal Pradesh using available maps, satellite images, aerial photographs and field data on different scales has already been prepared (Bhagat et al., 2004). The spatial and attribute database of the glaciers and glacial lakes of whole Himachal Pradesh was developed using remote sensing and geographic information system techniques. Major glaciers and glacial lakes were digitized on geo-coded LISS III satellite imagery using GIS software ILWIS 3.1. The spatial distribution of glaciers and glacial lakes was digitized on satellite image and verified from topographic maps. For the purpose, the state administrative boundary of Himachal Pradesh was divided into four major river basins viz., Beas, Ravi, Satluj and Chenab and four sub-basins which are not covered under these major river basins (Fig. 4). For the glacier inventory the study used the methodology developed by the Temporary Technical Secretary for the World Glacier Inventory (Muller *et al.,* 1977), and for the glacial lake inventory, the methodology developed by the Lanzhou Institute of Glaciology and Geocryology (LIGG) [LIGG/WECS/NEA 1988] was used with modification.

For the inventory of glaciers and glacial lakes in Himachal Pradesh, satellite imagery of Indian Remote Sensing Satellite Series 1 D (IRS1D) Linear Imaging and Self Scanning Sensor (LISS) III has been used. The methodology for the mapping and inventory of glaciers has been adopted from Mool *et al.*, (2001a and b). The inventory of glaciers has been based on topographic maps and satellite images. Glaciers were digitized on the satellite image and the identification, classification, and determination of stages of glaciers was accomplished by referring topographic maps of the glaciated regions of Himachal Pradesh. The spatial inventory is based entirely on topographic maps on a scale of 1:50,000 published in the 1960s to 1970s by the Survey of India. Altogether 110 topographic map sheets covered the whole of Himachal (Fig.5). The maps required for the study of the glaciers and glacial lakes fall within 60 sheets.

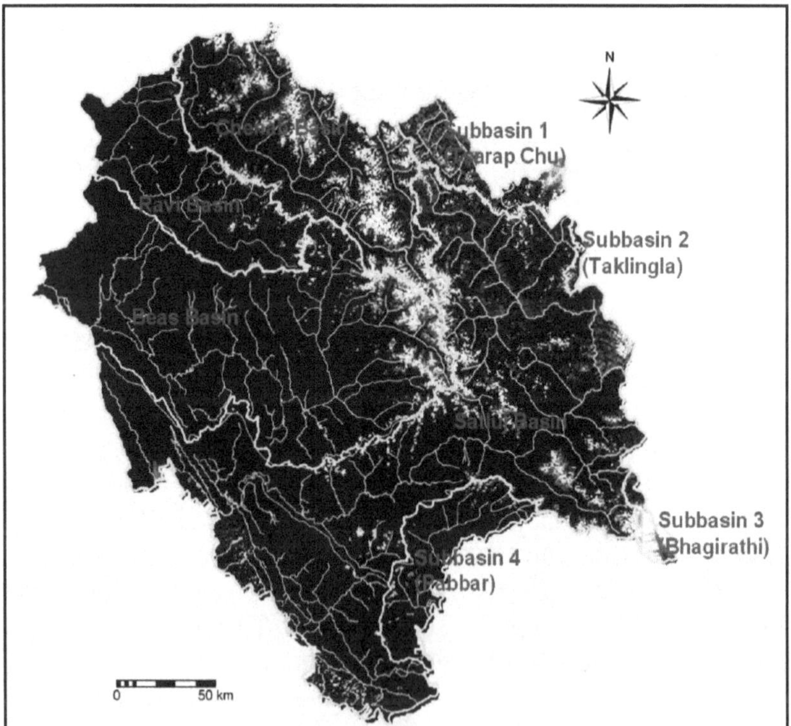

Figure 4. River network and basin wise distribution of glaciers in HP.

All the projection parameters of the topographic maps were incorporated in the images to make the prints compatible with the topographic maps. The coordinate/projection system parameters for Himachal Pradesh are as follows.

Projection	Albers Equal Area Conic
Ellipsoid	WGS 84
Datum	WGS 1984
False easting	0.0000000
False northing	0.0000000
Central meridian	82° 30′E
Central parallel	0° 0′ N
Latitude of first parallel	20° N
Latitude of second parallel	35° N

The minimum and maximum X and Y values required in the above geo-reference system in the Himachal area falling in Grid Zone II B are:

Min X,Y	:	-646838.500, 3298325.500
MAX X,Y	:	-328038.500, 3623225.500

To identify the physical conditions of individual glaciers and glacial lakes, different image enhancement techniques were used. Different colour composite (FCC) and individual spectral bands images highlighted different land-cover features using the knowledge of image interpretation keys: colour, tone, texture, pattern, association, shape, shadow etc. Fig. 6 shows colour composite images R3G4B2 (red to band 3, green to band 4, blue to band 2) and R4G3B2 of an IRS1D LISS3 of 19 September 1999 to 19 October 2001 of the Himachal Pradesh. In the colour composite images, one can identify different types of land cover, glaciers and glacial lakes. The glacier margins on each map are delineated and compared with satellite images, and the exact boundaries between glaciers and seasonal snow cover are determined. The coding system is based on the subordinate relation and direction of river progression according to the World Glacier Inventory. The classification of glaciers is adopted from the morphological classification of glaciers by the World Glacier Monitoring Service (WGMS) (Muller *et al.*, 1977).

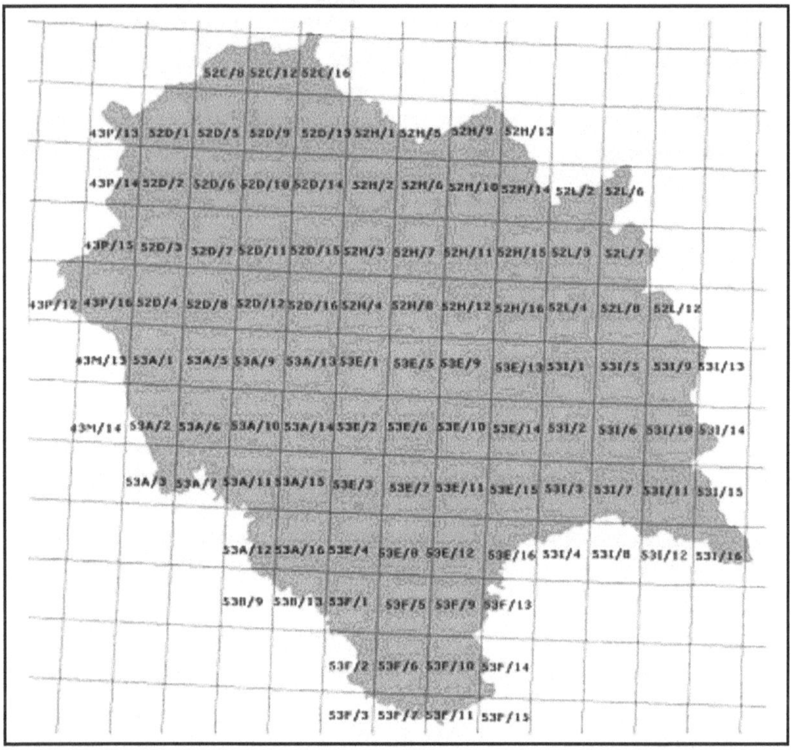

Figure 5. Index map for the 1:50,000 scale topographic maps of HP.

All the major river basins in Himachal Pradesh contained a significant number of glaciers. Basin wise distribution of glaciers including their number, area constituted and Ice reserve is shown in Table 1. Glaciers in this region generally occur above an elevation of 4,000 masl. It has been estimated that there are 2554 glaciers altogether within the territory of Himachal Pradesh, covering an area of 4160.58 sq. km with an ice reserves of 387.35 km^3 (Table 1 and Fig. 4). As evident from the Table 1, the Beas river basin has a total of 358 glaciers and covers an area of 758.18 sq. km with ice reserves of 76.42 km^3. The Ravi basin comprises of 198 glaciers. The total area covered by these glaciers is 235.21 sq. km with an ice reserve of 14.58 km^3. As many as 681 glaciers have been identified which feed Chenab river (Fig. 4).

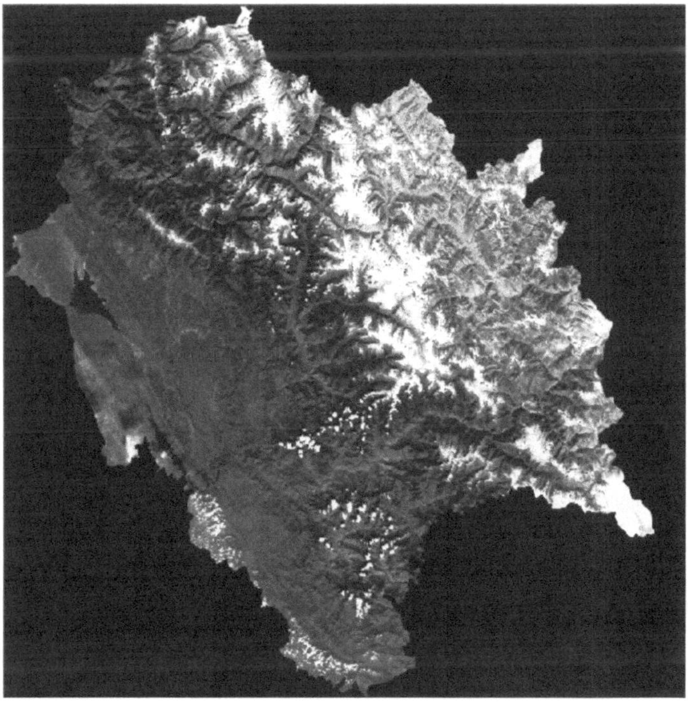

Figure 6. False Color Composite (FCC) of different spectral bands of IRS1D LISS3 of Himachal Pradesh region.

They cover an area of 1704.70 sq. km and an ice reserve of 187.66 km^3. The basin has also the credit of possessing the largest glacier of the state. The largest glacier, Bara Shigri glacier (mapped as Chenab_gr 585), is located at the south eastern end of the basin at latitude 32° 14' 58.83" N and longitude 77° 36' 52.13" E. It is a valley glacier about 34.24 km in length, constitutes an area of 180.26 sq. km and ice reserves of 43.53 km^3 (Fig. 7). Satluj is the longest and largest river traversing the Pradesh from east to west and plausibly, Satluj river basin is largest among five basins of the State. It spreads over 30° 22′ to 32° 42′ N latitude and 76° to 79° E longitude horizontally.

Eighty percent of its catchment is snow fed. It consists of 945 glaciers with a cumulative area of 1217.70 sq. km and an estimated ice reserve of 94.45 km^3. Although the number of glaciers in Satluj basin is comparatively high, the ice reserve and the area occupied is less than the Chenab basin.

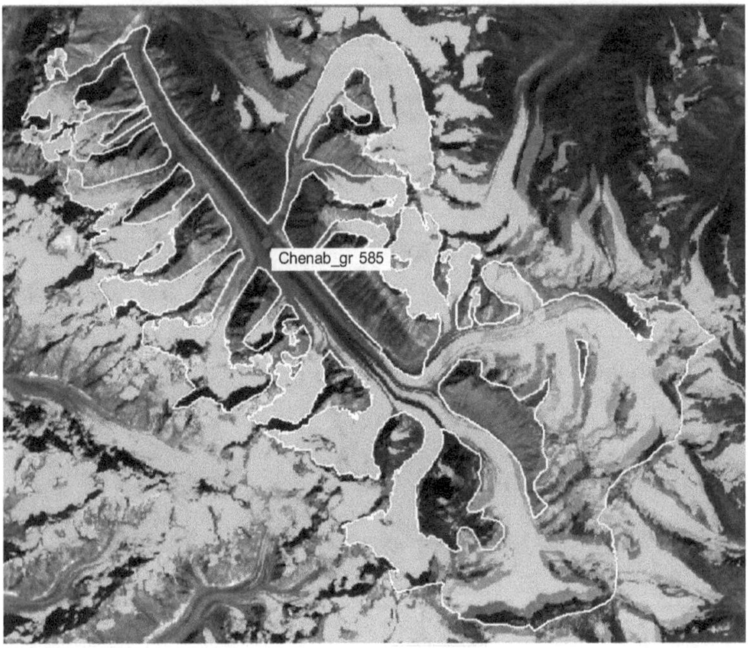

Figure 7. Bara Shigri glacier in the Chenab basin -one of the big glaciers in the world.

Sub basin 1 (Tsarap Chu) has 250 glaciers constituting an area of 163.33 sq.km and an ice reserve of 7.96 km^3. There are only 55 glaciers inventoried in Sub basin 2 (Taklingla) in an area of 32.04 sq.km and 1.38 km^3 ice reserve. Sub basin 3 (Bhagirathi) consists of 43 glaciers, 43.06 sq.km glaciated area and 2.43 km^3 ice reserves. The sub basin 4 (Pabbar) has 24 glaciers covering an area of 6.36 sq.km and an ice reserves of 0.19 km^3 and no lakes are observed in the sub basin.

The occurrence of glaciers has always been linked to climatic conditions. Climate is of fundamental importance to the inception and growth of glaciers. The form of the landscape dictates the threshold conditions for glacier occurrence and determines glacier morphology. In certain climatic conditions for glaciation, glaciers of different shapes and sizes are formed depending on the landscape. Mountain glacier regions are associated with climatic fronts, zones of maximum precipitation.

Table 1. Distribution of glaciers in Himachal Pradesh.

S.No.	Basin	Glacier Number	Area (km^2)	Ice reserve (km^3)
1	Beas	358	758.18	76.4
2	Ravi	198	235.21	16.88
3	Chenab	681	1704.7	187.66
4	Satluj	945	1217.7	94.45
5	Sub basin 1 (Tsarap Chu)	250	163.33	7.96
6	Sub basin 2 (Taklingla)	55	32.04	1.38
7	Sub basin 3 (Bhagirathi)	43	43.06	2.43
8	Sub basin 4 (Pabbar)	24	6.36	0.19
	Total	2554	4160.58	387.35

Generally, seven types of glaciers are observed in the Himachal Pradesh - mountain glaciers, mountain basin glacier, valley glaciers, cirque glaciers, niche glaciers, ice caps, and ice aprons (Table 2). Mountain glaciers are dominant in quantity and the profile shows a hanging nature. Other glaciers, except

Table 2. Different Types of Glaciers in Himachal Pradesh

	Number		Area		Ice reserve	
Glacier Type	Count	%	km^2	%	km^3	%
Cirque	155	71.67	23.716	22.84	0.52	15.94
Ice Cap	176	37.99	106.7	13.7	5.1	7.47
Mountain	116	48.23	22.2	37.69	0.61	42.67
Mountain basin	949	375.39	752.667	257.67	38.93	202.02
Niche	886	202.6	1778.94	310.68	128.19	305.65
Valley glacier	203	48.38	70.1	7.32	5.18	4.26

for valley glaciers, generally fall into the category of mountain glaciers but the thickness of ice is comparatively low. The number of valley glaciers is comparatively low but the corresponding areas and ice reserves are higher than those of mountain glaciers. The area and ice reserves of the individual valley glaciers are generally large owing to the fact that the ice thickness increases with increase in the area of the glacier.

Mountain glaciers are uncertain or miscellaneous, compound basins, compound basin or simple basin in the form of a hanging glacier. The major source of nourishment is snow and/or drift snow. Ice caps, cirque glaciers, niche glaciers, and ice aprons are other types of hanging mountain glaciers, but they are considered to be a different type due to their significance in size, shape, form, and ice thickness (Fig. 8). The most significant valley type glaciers are fewer in number and characterised by compound basins, compound basin and simple basin. They are mainly nourished by snow and drift snow at the headwater and by snow and ice avalanche at the lower valley. The adjoining part of the valley glacier at the headwaters is characteristically a mountain glacier, but due to its continuation into a valley glacier, the whole ice mass will be considered to be a valley glacier. Hence, the area of the valley glacier is higher than that of the mountain glaciers.

The longitudinal profile of the valley glacier from crown to toe shows an even or regular shape. As the headwater is steeper and has a gentle slope in the lower reaches, the profile makes the curve concave upwards. Due to the gentle slope at the lower reaches and the accumulation of debris derived from the headwater, glacial lakes develop in a supra glacial and moraine dammed form. Generally, the stability of glacial lakes is poor and there is always the chance of avalanches from mountain glaciers, which may break the damming material and cause glacial lake outburst floods (GLOFs).

Some important glaciers identified and mapped in different river basins of Himachal Pradesh are shown in Fig. 9.

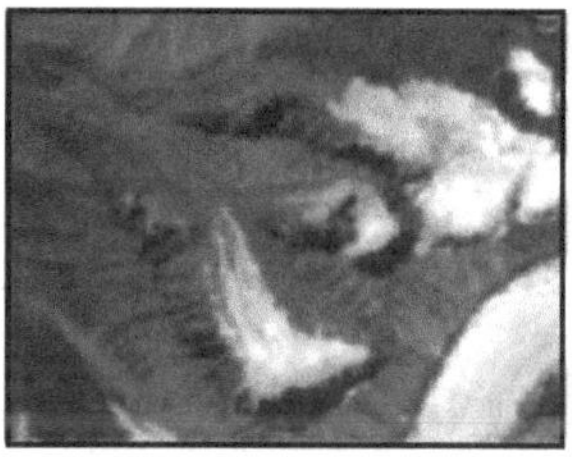

Cirque glacier mapped as Chenab_gr 580

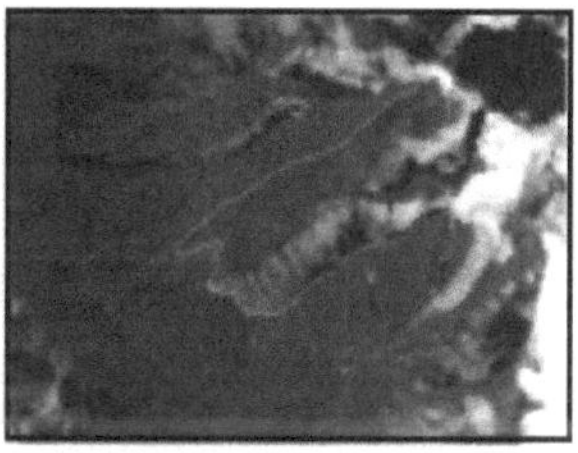

Ice Apron mapped as Chenab_gr 580

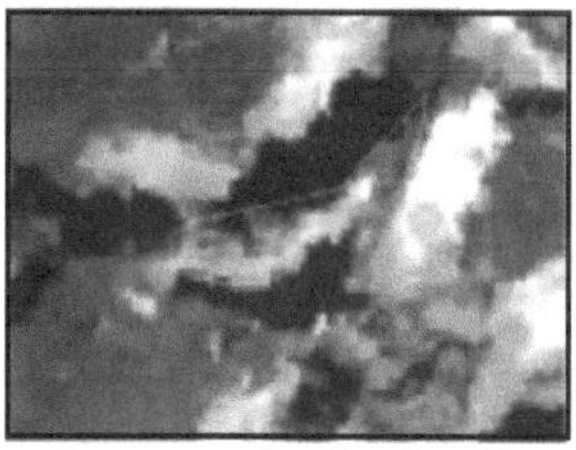

Ice cap mapped as Chenab_gr 32

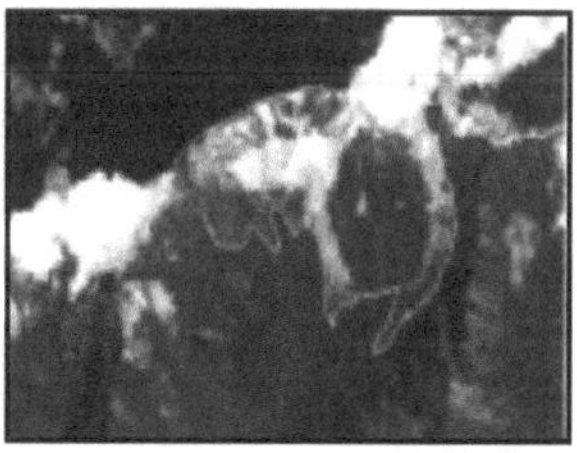

Mountain glacier mapped as Chenab_gr 81

Mountain basin mapped as Chenab_gr 429

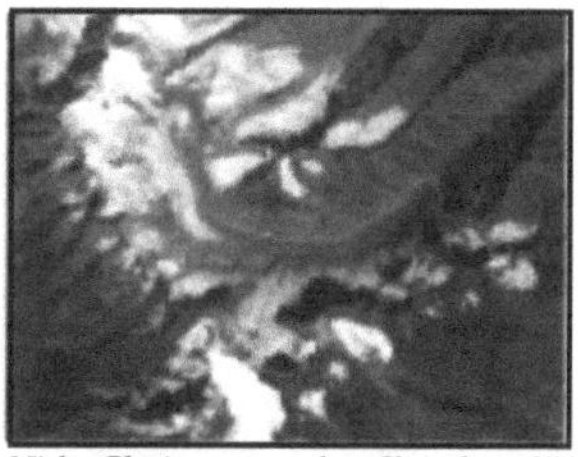

Niche Glacier mapped as Chenab_gr 256
identified as Chhangang palu lacier

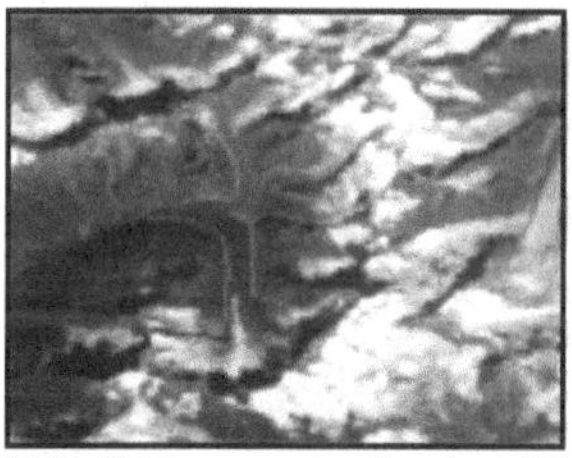

Valley Galcier mapped as Chenab_gr 203
identified as Gangs tong glacier

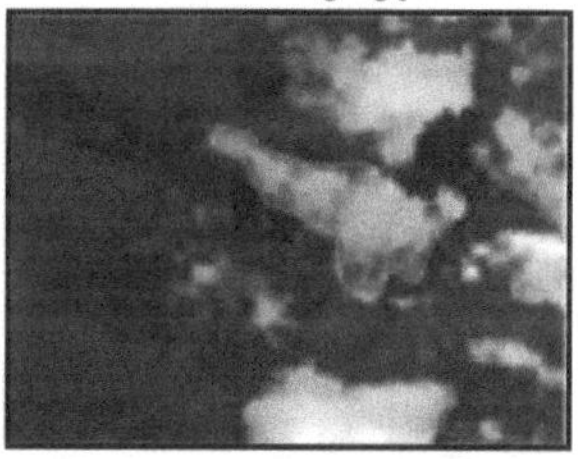

Cirque glacier mapped as Ravi_gr 184

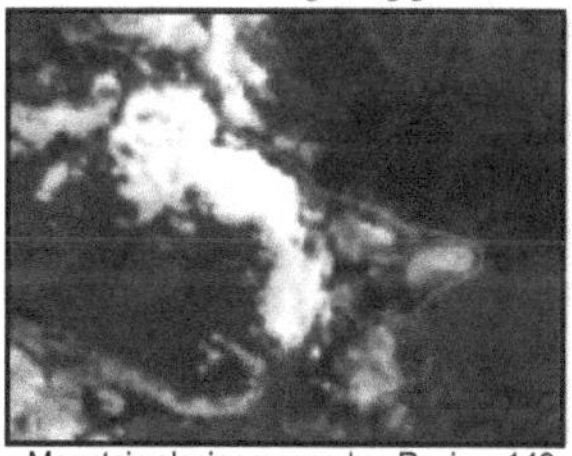

Mountain glacier mapped as Ravi_gr 149
identified as Guwala glacier

Mountain Basin mapped as Ravi_gr 123
identified as Badunag glacier

Figure 8. Different types of glaciers

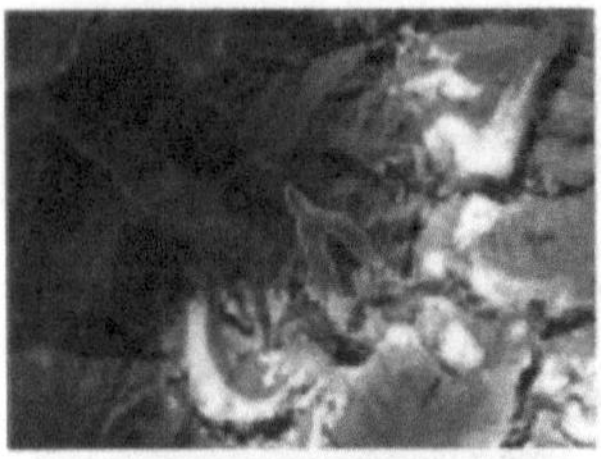

Saro Glacier mapped as satluj_gr 852 - Mountain basin

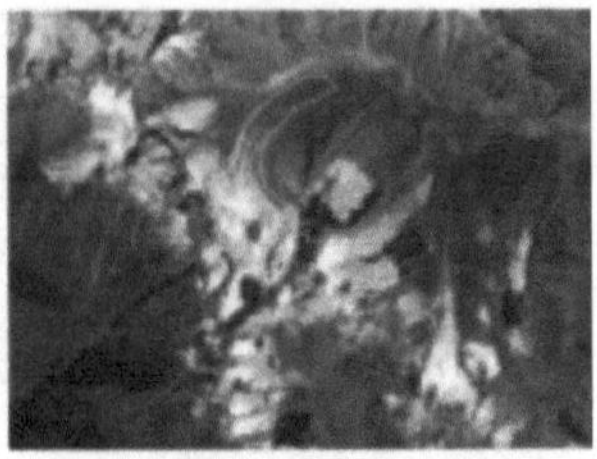

Sankari Glacier mapped as satluj_gr 100 -Mountain Basin

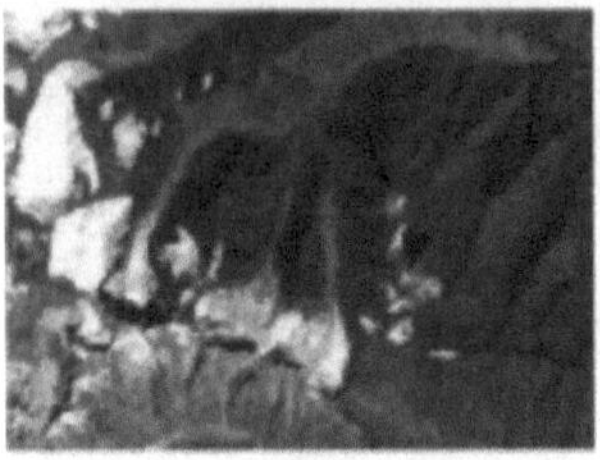

Phuldang Glacier mapped as satluj_gr 109 - Mountain Glacier

Chhota Shigri glacier mapped as Chenab_gr 587 - Mountain Glacier

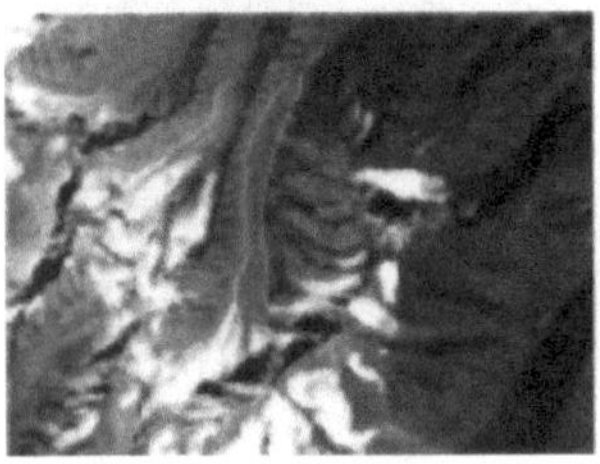

Lambar Glacier mapped as satluj_gr 844 - Valley Glacier

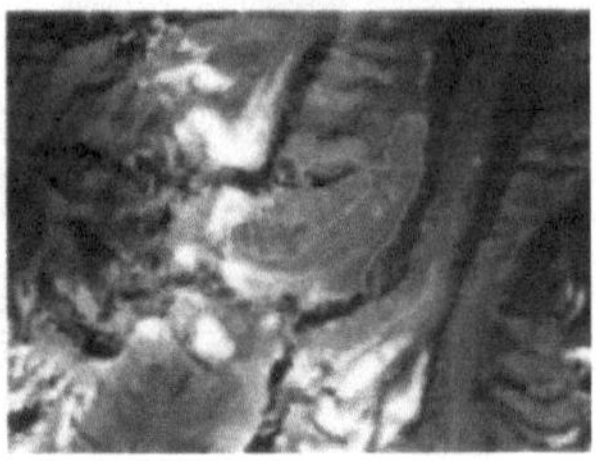

Garo Glacier mapped as satluj_gr 849 - Valley Glacier

Figure 9. Some of the important glaciers mapped and identified using SOI topomaps in Himachal Pradesh glacier inventory.

Glacial Lakes of Himachal Pradesh

A glacial lake is defined as a water mass existing in a sufficient amount and extending with a free surface in, under, beside and/or in front of a glacier and originated by glacier activities such as the retreating processes of a glacier. The lakes associated with perennial snow and ice, originate from glaciers. However, the isolated lakes found in the mountains and valleys far from the glaciers may not have a glacial origin. Due to the rapid rate of ice and snow melt, possibly caused by global warming, accumulation of water in these lakes has been increasing rapidly.

A glacial lake inventory for Himachal Pradesh has been systematically compiled for the drainage basins on the basis of topographic maps and satellite images calculating their area, length, width, depth, orientation, altitude etc (Bhagat *et al.* 2004). The lakes have been classified into erosion lakes, valley trough lakes, cirque lakes, blocked lakes, lateral and end moraine-dammed lakes, and supraglacial lakes. Different types of glacial lakes identified in the Chenab basin are presented in Fig 10.

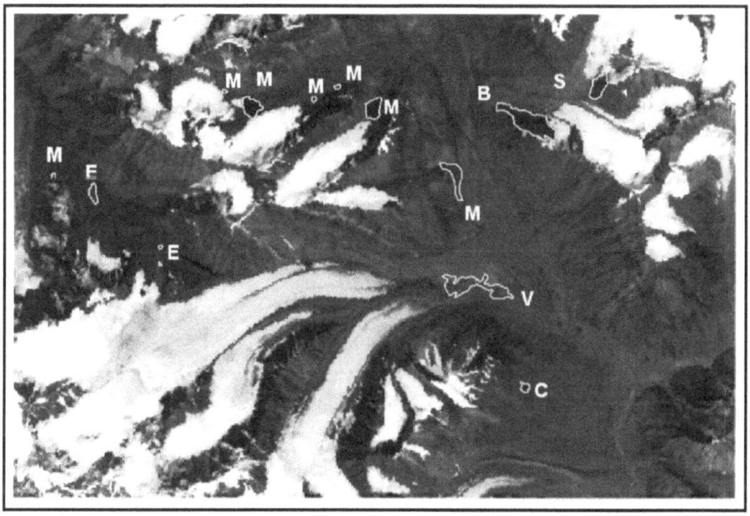

Figure 10. Lakes in the Chenab Basin showing the type of lakes. S-Supraglacial lake, M-Moraine Dammed Lake, E-Erosion Lake, B-Blocked Lake, C-Cirque Lake.

Altogether 156 lakes have been identified, which are distributed over all the basins in Himachal Pradesh, with a cumulative area of 385.22 km^2 (Table 3 and Fig 11). The lakes associated with glaciers were found out to be 141 in number. Fifty nine lakes have been identified in the Beas river basin, out of that 53 are found to be associated with glaciers hence could be classified as glacial lakes. A total of 17 lakes have been identified in the Ravi basin and 16 of them are associated with glaciers. The Chenab basin contains 31 glacial lakes (out of 33 in total) and out of the 40 lakes digitized in the Satluj basin, 35 are found to be associated with glaciers. No lake has been observed in the Sub basins 1 and 4, however, Sub basin 2 has six lakes and 3 had just one lake.

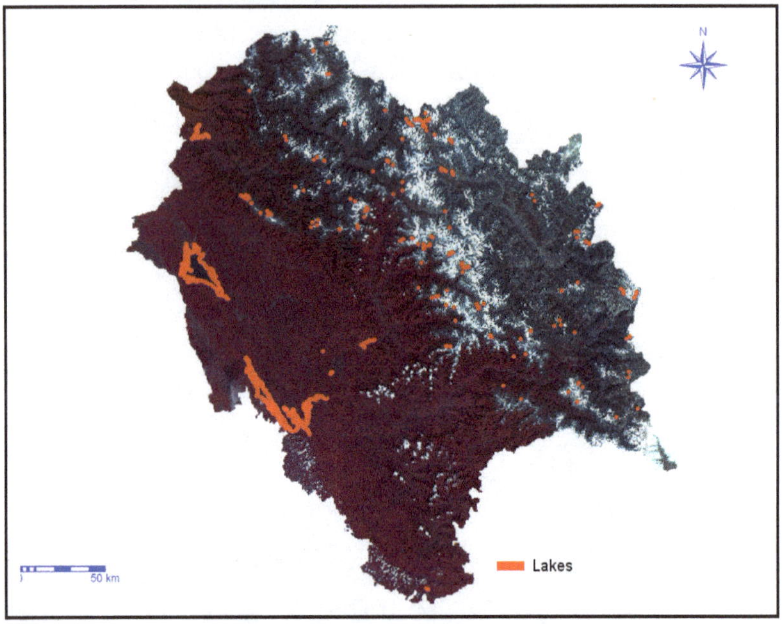

Figure 11. Lakes of Himachal Pradesh

Table 3. Distribution of lakes in the basins of Himachal Pradesh.

S.No.	Basin	Lake Number	Area (km²)
1	Beas	59	236.20
2	Ravi	17	9.16
3	Chenab	33	3.22
4	Satluj	40	136.46
5	Sub basin 1 (Tsarap Chu)	0	-
6	Sub basin 2 (Takling la)	6	0.16
7	Sub basin 3 (Bhagirathi)	1	0.02
8	Sub basin 4 (Pabbar)	0	-
	Total	156	385.22

In Himachal Pradesh almost all types of lakes have been found distributed over all the river basins (Table 4). Among these lakes, moraine-dammed lakes and blocked lakes are susceptible to breach out easily due to different phenomena.

Supraglacial lakes, when they start merging with one another to form a larger lake and finally change into a moraine-dammed lake, become dangerous.

Table 4. Types of lakes in Himachal Pradesh

Type	Number		Area (m^2)	
	Number	%	Area	%
Blocked	11	57.82	248779.4	39.85
Cirque	7	23.18	118333.9	1.08
Erosion	25	74.72	473307.4	3.69
Lateral Moraine	3	5.08	83066.65	0.04
Moraine dammed	60	186.92	1804502	130.89
Supraglacial	27	77.61	414060.3	11.24
Valley	23	174.63	373025836.6	403.21

Potentially Dangerous Glacial Lakes

Periodic or occasional release of large amounts of stored water results in a catastrophic outburst flood and is widely referred to as a jokulhlaup (Iceland), a debacle (French), an aluvión (South America), or a Glacial Lake Outburst Flood (GLOF) (Himalayas). As has been explained earlier, GLOF is a catastrophic discharge of water under pressure from a glacier. The potentially dangerous glacial lakes were identified and shown in Table 5 with the help of remote sensing and logical calculation in the geographic information systems (Mool, et al., 2001a and Mool, *et al.*, 2001b).

Moraine-dammed glacial lakes, which are still in contact or very near to the glaciers, are usually dangerous (Fig.12). These glacial lakes were partly formed between present-day glaciers and Little Ice Age moraine. The depositions of Little Ice Age moraines are usually about 300 years old, form high and narrow arch-shaped ridges usually with a height of 20–150 m, and often contain dead glacier ice layers beneath them. These end moraines are loose and unstable in nature. The advance and retreat of the glacier affect the hydrology between the present-day glacier and the lake dammed by the moraines.

Sudden natural phenomena with a direct effect on a lake, like ice avalanches or rock and lateral moraine material collapsing on a lake, cause moraine breaches with subsequent lake outburst events. Such phenomena have been well known in the past in several cases of moraine-dammed lakes, although the mechanisms at play are not fully understood.

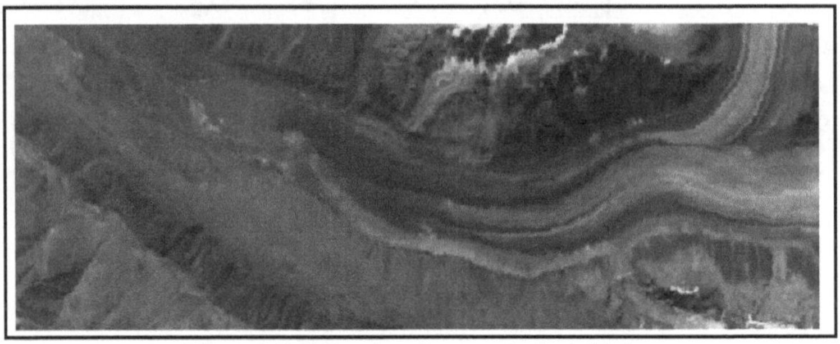

Figure 12. Moraine dammed glacial lake.

The criteria for identifying the potentially dangerous glacial lakes are based on field observations, processes and records of past events, geomorphic and geotechnical characteristics of the lake and surroundings, and other physical conditions as described in methodology section. For identification of potentially dangerous glacial lakes, the glacial lakes associated with glaciers such as supraglacial, valley, cirque and /or dammed by lateral moraine or end moraine with an area larger than 0.02 km^2 have been considered and they have been defined as major glacial lakes.

Based on different criteria, actively retreating glaciers and potentially dangerous lakes were identified using the developed spatial and attribute database complemented by multi - temporal remote sensing data sets. Sixteen glacial lakes have been identified (Table 5) in Himachal Pradesh as potentially dangerous lakes from the study of topographic maps, literature, and satellite images available (Bhagat et al., 2004). Among the potentially dangerous lakes, five lakes belong to the Beas basin, five lakes belong to the Chenab basin, one lake belong to the Ravi basin, three lakes belong to the Satluj basin and two belong to the sub-basin Takling La. Some of the potentially dangerous lakes as visible from satellite are shown in Fig. 13.

Table 5. Potentially dangerous glacial lakes in Himachal Pradesh.

S No	Lake Number	Name	Latitude Longitude	Length (m)	Area (m^2)
Beas Basin					
1	beas_gl 39		31°55'01.49"N, 77°31'51.84"E	237.1	20166
2	beas_gl 42		31°54'57.44"N, 77°31'08.62"E	178.2	22512
3	beas_gl 51		31°40'11.08"N, 77°37'13.93"E	449.8	43028
4	beas_gl 54		31°40'15.43"N, 77°35'57.62"E	410.5	24659
5	beas_gl 55		31°40'16.02"N, 77°35'31.67"E	311.2	27062
Satluj River Basin					
6	Satluj_gl 7		31°45'44.73"N, 78°06'44.25"E	262.1	27779
7	Satl uj_gl 10		32°00'37.86"N, 78°23'24.62"E	384.6	58659
8	Satluj_gl 13		32°15'57.63"N, 78°23'03.14"E	325.9	34504
Ravi River basin					
9	Ravi_gl 13		32°15'40.69"N, 76°44'24.84"E	278.5	57513
Chenab River Basin					
10	chenab_gl 7		32°50'05.82"N, 77°09'17.65"E	318.0	32804
11	chenab_gl 14		32°34'58.10"N, 77°11'15.66"E	1461.7	564940
12	chenab_gl 19		32°33'03.81"N, 77°31'26.00"E	2029.1	911043
13	chenab_gl 22		32°47'33.36"N, 77°22'32.24"E	534.2	119537
14	chenab_gl 25		32°47'31.43"N, 77°20'49.55"E	429.2	124631
Sub-basin2 (Takling la)					
15	Sub-basin2 1		32°12'31.98"N, 78°27'16.29"E	513.8	46676
16	Sub-basin2 2		32°13'05.47"N, 78°26'01.32"E	414.7	52793

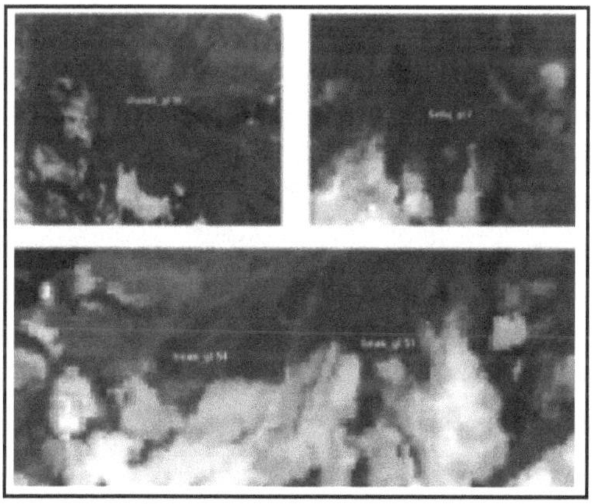

Figure 13. A few potential dangerous lakes identified in Himachal Pradesh.

Glacier Recession

Himachal glaciers have been in a state of general retreat since 1850. Though not in detail, a few glaciers in Himachal Pradesh have been studied by various workers. Retreats of the several glaciers based on the various literatures are also compiled in this chapter for the quick referencing.

Recent Indian head lines signify that the threat of disappearance looms large on many glaciers in Himachal Pradesh. Dr. Kulkarni and associates from Marine and Water Resources Group, Space Applications Centre, Ahmedabad are actively working on the effect of climatic variations on snow and glaciers in Himachal Pradesh. They have also prepared a glacier inventory of the Satluj Basin using remote sensing technique and monitored the glacier retreation in the basin using IRS PAN stereo data (Kulkarni, 1999) and observed that during the period 1963 -1997, Janapa Glacier retreated by 696 m, Shaune Garang by 923 m, Jorya Garang by 425 m, Naradu Garang by 550 m, Bilare Bange by 90 m, Karu Garang by 800 m and Baspa Bamak is retreated by 380 m. They further observed a massive glacial retreat of 6.8 km i.e. 178 m/year in Parbati glacier in Kullu District during 1962 to 2000. In these studies they observed an overall 19 per cent retreat in glaciated area and 23 per cent in glacier volume in last 39 years. These observations suggest that global warming has affected snow - glacier melt and runoff pattern in the Himalayas. Randhawa et al., (2001), using remote sensing techniques, prepared a glacier inventory for the Satluj and Beas basins in Himachal Pradesh under Himalayan Glacier Inventory program on the scale of 1:50,000. The mapping was done using LISS I satellite data. The study indicated the presence of total 334 glaciers and 1987 snow fields in the entire Satluj basin covering an area of 2697 sq. km as a whole. Of the 38 moraine dammed lakes identified in the entire basin, 14 lakes falls in the Himachal Himalayas where as the remaining lakes falls in the Tibet Himalayas in the Satluj catchment.

Glacier mass balance is an important glaciological parameter and is intimately related to climate. A few glaciers in the Baspa basin have been monitored for long term mass balance observation (Kulkarni et al., 2004) using accumulation area ratio. Mass balance was estimated during 2001 and 2002 for 19 glaciers in the basin, suggesting overall specific mass balance value of - 90 and - 78 cm, respectively. Their investigation suggested a loss of 0.2347 km^3 of glacial ice in the last two years. The investigation has shown that four glaciers have no accumulation area, as these are located in lower-altitude zones. These glaciers are expected to face terminal retreat due to lack of formation of new ice. This is likely to pose serious problem of availability of water to many villages located in the Baspa basin.

Geological Survey of India at Lucknow, as a part of their study on "Glacier recession in Himalayas" depicted that Ravi basin has 172 glaciers covering an area of 193 km^2, Chenab-1278 galcier with 3059 km^2 area, Beas-277 glaciers with 579 km^2 area and Satluj basin has 926 glaciers covering an area of 1252 km^2. They have also studied the Satluj basin glaciers Gara, Gor Garang, Shaune Garang, Nagpo Tokpo in detail and observed an average retreat of 4.22 -6.8 m/year in all these. In Chenab basin they studied Bara Shigri, Chhota Shigri, Miyar, Hamtah, Nagpo Tokpo, Trilokinath and Sonapani and concluded that the average glacier retreat rate ranges from 6.81 m/year for Chhota Shigri to 29.78 m/year for Bara Shigri glacier.

It is therefore important that the impacts of global warming on snow and glacier melting and hence water resources should be fully understood. Floods and droughts are other major events that dominate the headlines of newspapers in the region every year during monsoon and dry periods. Floods can include glacial lake outburst floods, landslide dammed flash floods, high rainfall-induced floods and floods in ephemeral rivers (which remain dry for almost eight months a year). The region has witnessed major climate-induced disasters in recent years, such as the consecutive catastrophic monsoon floods in Satluj basin during 1997 and 2000 (Sharma, 2000). It is difficult to definitely relate these events to climate change, the normal as well as abnormal variability in the climate would affect river runoff pattern.

Unlike the neighbouring countries Nepal, Tibet, Bhutan which have experienced enormous losses due to GLOFs, no major apparent GLOF event has ever been reported in the state, however, quite a few lakes have been identified which may pose danger in the coming future as the glaciers are retreating at an alarming rate due to global warming. There are significant number of lakes in all the basins having the area ranging between 0.1 and 1.25 sq. km. Among the glacial lakes studied, a lake associated with Geopang gath glacier in the Chenab basin has received the greatest attention.

Receding glaciers means decline in water storage as glaciers are storehouses of water. Global warming is definitely causing this. Besides, creation of glacial lakes and the hazards associated with it should also be monitored. A reduction in water from the mountains would further affect the economy of the region by limiting the energy from hydropower plants and hampering industrial productivity (Johannesson 1997). Power shortages in India are already at about 10 percent of total electrical energy and 20 percent of peak capacity requirements (World Bank, 2000).

Development of early warning systems along with regular monitoring of glaciers and glacial lakes in the Himachal Himalaya should be taken up on priority using Geographic Information System and Remote Sensing (GIS-RS). The entire State was plunged into a crisis due to a recent blockade of Pareechu tributary of Spiti river which drains into Satluj River. Until now, according to different news, "the 'dam' is intact but it is unstable." During the winter time in such terrain the lake generally freezes and will have false sense of security in downstream areas. Necessary mitigation measures must be taken care of in time to prevent the damage to lives and properties in the downstream. In such situation sharing of information and regional cooperation is very important.

It may be concluded that the present day risk for an outburst from glacial lakes occurring in Himachal Pradesh may be low, but the risk of outburst of glacial Lakes in the future could be anticipated high and it might occur in coming 15-20 years considering the present trend of climate change (Häuslar and Leber (2000). It is proposed that besides making a temporal inventory, a close monitoring of these lakes is required to assess the change in their behaviour. This will help in strengthening database and also help in undertaking an appropriate pre-disaster mitigation measure and in avoiding flash flood tragedies common in the hilly region. Development of an early warning system can be of great help in saving precious lives and property.

References

Ageta, Y. and Kadota, T. (1992) 'Predictions of Changes of Glacier Mass Balance in the Nepal Himalaya and Tibetan Plateau: A Case Study of Air Temperature Increase for Three Glaciers'. In *Annals of Glaciology*, 16: 89-94.

Arnoff, S. (1989) Geographic Information Systems: A Management Perspective. Ottawa, Canada: WDL

Bhagat, R.M., Vaibhav Kalia, Chitra Sood, Pradeep Mool and S.R Bajracharya (2004) Himachal Pradesh Himalaya - Inventory of glaciers and glacial lake outburst floods (GLOFs) affected by global warming in the mountains of Himalayan region. Technical report submitted to APN-START/UNEP, Japan.

Bhagat R.M., Vaibhav Kalia and Chitra Sood, Kamlesh Singh, Ranbir Rana and Anurag Saklani (2005) Satluj River Basin: A Preliminary investigation into the water reserve mapping, assessment and viability of micro/mega hydropower projects and suggesting alternatives. Technical report submitted to Ministry of Power, Govt. of Himachal Pradesh.

Cahohai, Liu; Liangfu, Ding (1986) 'The Newly Progress of Glaciers Inventory in Tianshan Mountains'. *Journal of Glaciology and Geocryology*, 8(2): 168-169.

Costa, J.E. (1985) *Floods From Dam Failures,* Open File Report. USA:US Geological Survey.

Dwivedi K. (2000) The Tam Pokahri Glacier Lake Outburst Flood Event, Disaster Review, Series VII, July 2000, DWIDP/DMSP/JICA

Fujita, K., Nakawo, M., Fujii, Y., and Paudyal, P. (1997). Changes in glaciers in Hidden Valley, Mukut Himal, Nepal Himalayas, from 1974 to 1994. *Journal of Glaciology* 43, 583-588.

Haeberli, W. (1990) Glacier and permafrost signals of 20th century warming. *Annals of Glaciology*, 14, 99-101.

Häusler, H.; Leber, D.; Schreilechner, M.; Morawetz, R.; Lentz, H.; Skuk, St.; Meyer, M.; Janda, Ch.; Burgschwaiger, E. (2000) Final Report of Raphstreng Tsho Outburt Flood Mitigatory Project (Lunana; Northwestern Bhutan): Phase II. Vienna, Austria: Institute of Geology, University of Vienna

IPCC, (2001) Climate Change 2001: Impacts, Adaptation and Vulnerability. Contribution of Working Group II to the Third Assessment Report of the Intergovernmental Panel on Climate Change. James J McCarthy, Osvaldo F Canziani, Neil A Leary, David J Dokken, Kasey S White; editors. Cambridge University Press. United Kingdom. 533-590 pp.

IPPC, (1996) Climate change 1995, The science of climate change. Cambridge University Press, pp. 572.

Jain, S. (1991) Standing up for the trees: women's role in Chipko movement. Sally Sontheimer (Ed.). Women and the environment: a reader. New York: Monthly Review Press.

Johannesson, T. (1997) The Response of Two Icelandic Glaciers to Climate Warming computed with a Degree-Day Glacier Mass Balance Model coupled to a Dynamic Glacier Model, *Journal of. Glaciology*, 43(144): 321-327.

Kayastha, S.L. (1971) "Himachal Region" In R.L. Singh (Ed.). India-A Regional Geography. National Geographical Society of India, Varanasi, pp 390-442.

Kulkarini, A.V., P. Mathur, B.P. Rathore, S. Alex, N. Thakur, M. Kumar (2002) Effect of global warming on snow ablation pattern in the Himalaya. *Current Science*, 83, 2: 120-123.

Kulkarni A.V. (1999) Moraine dammed glacial lake studies using remote sensing techniques. *Himalayan Geology*. 17:161-164

Kulkarni, A. V. Bahuguna I.M. (2002) Glacier retreat in the Baspa basin, Himalaya, monitored with satellite stereo data. *Journal of Glaci*ology, 48(160):171-172.

Kulkarni, A.V. (2003) Effect of climatic variations on snow and glaciers in Himachal Pradesh. In Proceedings of the two days brain storming session on mountain environment & climate change over Himalayan Region at Shimla, 14-15th march 2003.

Kulkarni, A.V., Rathore, B.P. and Alex, S. (2004) Monitoring of glacial mass balance in the Baspa basin using accumulation area ratio method. *Current Science*, 86:185-190.

LIGG/WECS/NEA (1988) Report on First Expedition to Glaciers and Glacier Lakes in the Pumqu (Arun) and Poique (Bhote-Sun Kosi) River Basins, Xizang (Tibet), China, Sino-Nepalese Investigation of Glacier Lake Outburst Floods in the Himalaya. Beijing, China: Science Press.

Mayewski, P.A. & Jeschke, P.A. (1979) Himalayan and trans-Himalayan glacier fluctuations since AD 1812. Arctic and Alpine Research 11(3), 267-287.

Mool, P.K.; Bajracharya, S.R.; and Joshi, S. P. (2001a) Inventory of Glaciers, Glacial Lakes, and Glacial Lake Outburst Flood Monitoring and Early Warning System in the Hindu Kush-Himalayan Region, Nepal. 364P. ICIMOD in cooperation with UNEP/RRC-AP, ISBN 92 9115 331 1, Published by ICIMOD, Kathmandu, Nepal.

Mool, P.K.; Wangda, D.; Bajracharya, S.R.; Joshi, S. P.; Kunzang, K.; and Gurung, D.R. (2001b) Inventory of Glaciers, Glacial Lakes, and Glacial Lake Outburst Flood Monitoring and Early Warning System in the Hindu Kush-Himalayan Region, Bhutan. 227. ICIMOD in cooperation with UNEP/RRC-AP, ISBN 92 9115 345 1, Published by ICIMOD, Kathmandu, Nepal.

Muller, F., Caflish, T., Muller, G. (1977) Instruction for Compilation and Assemblage of Data for a World Glacier Inventory. Zurich: Temporary Technical Secretariat for World Glacier Inventory, Swiss Federal Institute of Technology, Zurich

Negi, S.S. (1991). Himalayan rivers, lakes and glaciers. Indus Publishing Co. New Delhi, 182 p.

New Scientist, (1999) "Flooded Out: Retreating glaciers spell disaster for valley communities", 5th June, 1999, p. 18.

New Scientist, (2003) "Glacier crack places Peruvian city in peril", 14th April, 2003.

Randhawa, S.S., Sood, R.K. and Kulkarni, A.V. (2001) Delineation of moraine - dammed lakes in Himachal Pradesh using high resolution IRS - LISS III satellite data.

Sharma, A. (2000) Over 150 feared killed in Himachal flash-floods. The Indian Express, August 2 & 3, 2000.

Sharma, S., (2001) The Ganges Crisis – Government lacks strategic plan. In Proc.: Bonn 2001- International Conference on Freshwater, Bonn, Germany, 3- 7, December 2001.

Singh, P., Jain, S.K., Kumar, N. and Singh, U.K. (1994a) Snow and Glacier Contribution in the Chenab Rriver at Akhnoor, CS (AR)-131, National Institute of Hydrology, Roorkee, India.

Singh, P., Jain, S.K., Kumar, N. and Singh, U.K. (1994b) Snow and Glacier Contribution in the Ganga river at Devprayag, CS (AR)-132, National Institute of Hydrology, Roorkee, India.

Wanchang, Z., Ogawa, K., Besheng, Y. and Yamaguchi, Y. (2000) A monthly stream flow model for estimating the potential changes of river runoff on the projected global warming, *Hydrological Processes*, 14, 1851-1868.

Wood, F.B (1990) Monitoring global climatic change: the case of greenhouse warming. *Bulletin of the American Meteorological Society*, 71, 1:42 –52.

World Bank, (2000) The World Bank Group – India Country Brief 2000.

Yamada, T. (1996) Preliminary report on glacier lake outburst flood in the Nepal Himalayas. Report 4/1/291191/1/1 Seq. No. 387, *Water & Energy Secretariat*, HMG, Kathmandu, Nepal.

❑❑❑

GEOINFORMATICS APPLICATIONS IN AGRICULTURE, 2007
Eds. : A.K. Singh and U.K. Chopra; pp. 175-207
New India Publishing Agency, New Delhi (India)
E-mail : newindiapublishingagency@gmail.com
Web: www.bookfactoryindia.com

7

Monitoring and Management of Water Resources

Sanjay K. Jain, Rakesh Kumar & K. D. Sharma

Key words : GPS, GIS, DSS, SDSS, RS, Digital Elevation Model (DEM), Ground Water Exploration, Conjunctive Use of Surface and Ground Water, Digital Terrain Modeling (DTM).

Introduction

Water is one of the most precious gifts of nature to living beings and particularly to the mankind. On account of erratic distribution of rainfall both in time and space, the country is subjected to cycles of floods and droughts. With the increase in demand for water for various purposes and to mitigate the twin problems of floods and droughts faced by the country, a coordinated approach for optimum utilization and proper management of this precious resource is of prime importance. In addition, proper development and management of both irrigation and drainage systems are essential. This would require integrated management of surface water as well as gorund water. All the water related problems whether mitigating droughts or floods or better irrigation management practices, dictate for adoption of state of the art water management practices. Water management deals with the control, distribution and allocation of water flows and the treatment of rivers and catchments, consequently, solutions to many of the problems in water management require the knowledge and expertise from diverse sources.

India receives annual precipitation of about 4000 km^3, including snowfall. Out of this, monsoon rainfall is of the order of 3000 km^3. Rainfall in India is dependent on the South-West and North-East monsoons, on shallow cyclonic depressions and disturbances and on local storms. Most of it takes place under the influence of South-West monsoon between June to September except in Tamil Nadu, where it is under the influence of North-East monsoon during October and November. India is gifted with a river system comprising of more than 20 major rivers with several tributaries. Many of these rivers are perennial and some of these are seasonal. The rivers like Ganges, Brahmaputra and Indus originate from the Himalayas and carry water throughout the year. The snow and ice melt of the Himalayas and the base flow contribute the flows during the lean season. Lal (2001) mentioned that more than 50% of water resources of India are located in various tributaries of these river systems. Average water yield per unit area of the Himalayan rivers is almost double that of the south peninsular rivers system indicating the importance of snow and glacier melt contribution from the high mountains. Apart from the water available in the various rivers of the country, the ground water is also an important source of water for drinking, irrigation and industrial uses etc. It accounts for about 80% of domestic water requirement and more than 45% of the total irrigation in the country. As per the international norms, if per-capita water availability is less than 1700 m^3 per year then the country is categorized as water stressed and if is less than 1000 m^3 per capita per year then the country is classified as water scarce. In India per capita surface water availability in the year 1991 and 2001 were 2309 and 1902 m^3 and these are projected to reduce to 1401 and 1191 m^3 by the years 2025 and 2050, respectively. The increasing water scarcity combined with its deteriorating quality, will have far reaching impact on human health and socio-economic condition. Therefore, there is an urgent need for development of water resources and management strategies, which includes rain water harvesting, flood and drought management, water conservation and ground water recharging in low rainfall areas or transfer of water through network of canals/rivers from water surplus regions to water deficit regions. This calls for a scientific database on hydrology both spatial and non-spatial types for critical evaluation of the existing situation in terms of water resources status and for proper decision makings. Remote sensing data provide valuable information for many water resources management and monitoring tasks, which will facilitate sustainable management of water resources. Remote sensing in combination with Global Positioning System (GPS) and GIS can produce the terrain maps with high lacational accuracy containing detailed information on the variables under study. Decision Support System (DSS) have been evolved from practices in the management of information systems, particularly in the field of data processing in business sciences. A DSS can be defined as an interactive computer based system, which permits a combination of knowledge sources from various domains in order to help decision makers to solve complex problems. When applied to water management, DSS requires a spatial dimension and is therefore usually incorporated in a GIS, thus forming a Spatial Decision Support System (SDSS). Currently available DSS and SDSS are being used for decision making in water resources planning and management.

Before discussing different aspects of water resources/ management and application of remote sensing/GIS technologies, water resources of India have been described in the following section.

Water Resources of India

Although, India occupies only 3.29 million km^2 geographical area, which forms 2.4% of the world's land area, it supports over 15% of the world's population. The population of India as on 1st March 2001 stood at 1,027,015,247. Thus, India supports about 1/6th of world population, 1/50th of world's land and 1/25th of world's water resources (Water Management Forum, 2003). India also has a livestock population of 500 million, which is about 20% of the world's total livestock population. More than half of these are cattle, forming the backbone of Indian agriculture. The total utilizable water resources of the country are assessed as 1086 km^3. A brief description of surface and groundwater water resources of India is given ahead.

Surface water resources

In the past, several organizations and individuals have estimated water availability for the nation. Recently, the National Commission for Integrated Water Resources Development (IWRS, 1998) estimated the basin-wise average annual flow in Indian river systems as 1953 km^3. The details are given in Table 1.

Table 1. Basinwise average flow and utilizable water (NCIWRD, 1999) (in km^3/year)

Sl. No.	River Basin	Average Annual Flow	Utilizable Flow
1	Indus	73.31	46
2	Ganga-Brahmaputra-Meghna Basin		
	2a Ganga	525.02	250
	2b Brahmaputra sub-basin	629.05	24
	2c Meghna (Barak) sub-basin	48.36	
3	Subarnarekha	12.37	6.81
4	Brahmni-Baitarani	28.48	18.3
5	Mahanadi	66.88	49.99
6	Godavari	110.54	76.3
7	Krishna	69.81	58
8	Pennar	6.32	6.86
9	Cauvery	21.36	19
10	Tapi	14.88	14.5
11	Narmada	45.64	34.5
12	Mahi	11.02	3.1
13	Sabarmati	3.81	1.93
14	West flowing rivers of Kachchh and Saurashtra including Luni	15.1	14.98
15	West flowing rivers south of Tapi	200.94	36.21
16	East flowing rivers between Mahanadi and Godavari	17.08	
17	East flowing rivers Between Godavari and Krishna	1.81	13.11
18	East flowing rivers Between Krishna and Pennar	3.63	
19	East flowing rivers Between Pennar and Cauvery	9.98	16.73
20	East flowing rivers south of Cauvery	6.48	
21	Area of North Ladakh not draining into Indus	0	NA
22	Rivers draining into Bangladesh	8.57	NA
23	Rivers draining into Myanmar	22.43	NA
24	Drainage areas of Andman, Nicobar and Lakshadweep Islands	0	NA
	Total (Rounded)	**1953**	**690**

Utilizable water resource is the quantum of withdrawable water from its place of occurrence. Within the limitations of physiographic conditions and socio-political environment, legal and constitutional constraints and the technology of development available at present, utilizable quantity of water from the surface flow has been assessed by various authorities differently. The utilizable annual surface water of the country is 690 km^3 as given in Table 1 (NCIWRD, 1999). There is considerable scope for increasing the utilization of water in Ganga-Brahmaputra basins by construction of storages at suitable locations in neighbouring countries.

Ground water resources

The annual potential natural ground water recharge from rainfall in India is about 342.43 km^3, which is 8.56% of total annual rainfall of the country. The annual potential ground water recharge augmentation from canal irrigation system is about 89.46 km^3. Thus, total replenishable groundwater resource of the country is assessed as 431.89%. After allotting 15% of this quantity for drinking, and 6 km^3 for industrial purposes, the remaining can be utilized for irrigation purposes. Thus, the available ground water resource for irrigation is 361 km^3 of which utilizable quantity (30%) is 325 km^3. The estimates by Central Ground Water Board (CGWB) of total replenishable ground water resource, provision for domestic, industrial and irrigation uses and utilizable ground water resources for future use are given in Table 2. The basin wise per capita water availability varies between 13,393 m^3 per annum for Brahmaputra-Barak basin to about 300 m^3 per annum for Sabarmati basin. The basin-wise ground water potential of the country is given in Table 3.

Water Resources Management in India

Water is a most vital natural resource and is under tremendous stress due to exponential growth in population, rapid industrialization and urbanisation in our country. The increasing human and livestock population has reduced the per capita water availability significantly over a period of time

Table 2. Ground water resources of India (in km³)

1.	Total Replenishable Ground Water Resource	432
2.	Provision for Domestic, Industrial and other Uses	71
3.	Ava ilable Ground Water Resource for Irrigation	361
4.	Utilizable Ground Water Resource for Irrigation (90% of the Sl. No. 3)	325
5.	Total Utilizable Ground Water Resource (Sum of Sl No. 2 and 4)	396

Table 3. Ground water potential in river basins of India (IWRS, 1998) (Pro rata basis) (in km³/year)

Sl. No.	Name of the basin	Total replenis-able ground water resources	Provision for domestic, industrial and other uses	Available ground water for irrigation	Net draft	Balance ground water potential	Level of ground water development (%)
1.	Brahmani with Baitarni	4.05	0.61	3.44	0.29	3.16	8.45
2.	Brahmaputra	26.55	3.98	22.56	0.76	21.80	3.37
3.	Chambal Composite	7.19	1.08	6.11	2.45	3.66	40.09
4.	Cauvery	12.30	1.84	10.45	5.78	4.67	55.33
5.	Ganga	170.99	26.03	144.96	48.59	96.37	33.52
6.	Godavari	40.65	9.66	30.99	6.05	24.94	19.53
7.	Indus	26.49	3.05	23.43	18.21	5.22	77.71
8.	Krishna	26.41	5.58	20.83	6.33	14.50	30.39
9.	Kutch & Saurashtra Composite	11.23	1.74	9.49	4.85	4.64	51.14
10.	Madras and South Tamil Nadu	18.22	2.73	15.48	8.93	6.55	57.68
11.	Mahanadi	16.46	2.47	13.99	0.97	13.02	6.95
12.	Meghna	8.52	1.28	7.24	0.29	6.95	3.94
13.	Narmada	10.83	1.65	9.17	1.99	7.18	21.74
14.	Northeast Composite	18.84	2.83	16.02	2.76	13.26	17.20
15.	Pennar	4.93	0.74	4.19	1.53	2.66	36.60
16.	Subarnarekha	1.82	0.27	1.55	0.15	1.40	9.57
17.	Tapi	8.27	2.34	5.93	1.96	3.97	33.05
18.	Western Ghat	17.69	3.19	14.50	3.32	11.18	22.88
	Total	**431.43**	**71.08**	**360.35**	**115.21**	**245.13**	**31.97**

and thus put the available water under severe stress. As per a World bank study, the per capita availability of water in India has decreased from 5000 cubic meter per year in 1947 to less than 2000 cubic meter in 2002 and expected to drop further to less than 1500 cubic meter in 2025. Today, water scarcity is the biggest challenge facing the country (Sinha et al., 2004). Therefore, there is an urgent need for water resources development and management in India.

For planning, design and operation of the water resources projects, temporal & spatial data of various hydrometeorological variables as well as basin characteristics are required. Surveying and mapping is basic to effective water management. In the conventional engineering studies, the traditional inputs are topographical maps, aerial photographs and ground survey. Topographical maps depicting current status of the region under study are seldom available because of longer revision cycles. Aerial photography is expensive besides being procedurally cumbersome. Ground surveys are too costly, time consuming and laborious. It is here that remote sensing and Geographic Information System (GIS) help in creating an appropriate information base for efficient management of the water resources. The synoptic view provided by satellite remote sensing and the analysis capability provided by GIS offer a technologically appropriate method for studying various features related to land and water resources. It is worth mentioning here that remote sensing cannot entirely do away with conventional surveying, but it can limit the ground surveys.

Remote sensing is a tool permitting accurate and real time evaluation, continuous monitoring or surveillance and forecast of inland water resources. Remote sensing systems are used to observe the earth's surface from different platforms such as satellites and aircrafts, and make it possible to collect and analyse information about resources and environment over large areas. Remote sensors record electromagnetic energy reflected or emitted from earth's surface. Different kinds of objects or features such as soils, vegetation and water reflect and emit energy differently. This characteristic makes it possible to measure, map and monitor these objects and features using satellite or aircraft borne remote sensing systems. Satellite images provide a low cost and potentially rapid means to monitor and map the different land cover features. One of the greatest advantages of using remote sensing data for hydrological monitoring and modeling is its capability to generate information in spatial and temporal domain, which is important in water management studies. Increasing spatial, spectral and temporal resolution of data from satellite sensors has led to a quantum jump in our capabilities for remote sensing applications towards water resources management.

For many water related studies, remote-sensing data alone are not sufficient and they have to be merged with the data from other sources. Hence, a multitude of spatially related (i.e. climatic and geographic) data concerning rainfall, evaporation, vegetation, geomorphology and soils has to be considered. In addition, information is also required such as locations and types of tube wells, rain and river gauges etc. Thus, the fast storage, retrieval, displays and updating of map contents consitute important functions. A system that can store the data, select and classify the stations and perform mathematical and sorting operations is called a database and information can be extracted from it for a given purpose. If this information can also be displayed in the form of maps, we can speak of geographic information. So this complete set of information forms the Geographic Information System (GIS). The GIS is an effective tool for storing, managing, and displaying spatial data often encountered in hydrology and water resources management related studies. The GIS technology integrates common database operations, such as queries and statistical analysis, with the unique visualization and geographic analysis benefits offered by maps and spatial databases. The use of remote sensing data in GIS is shown in Figure 1.

Problems associated with water resources development and efficient utilization are varied and too many in nature like frequent floods and droughts, waterlogging and salinity in

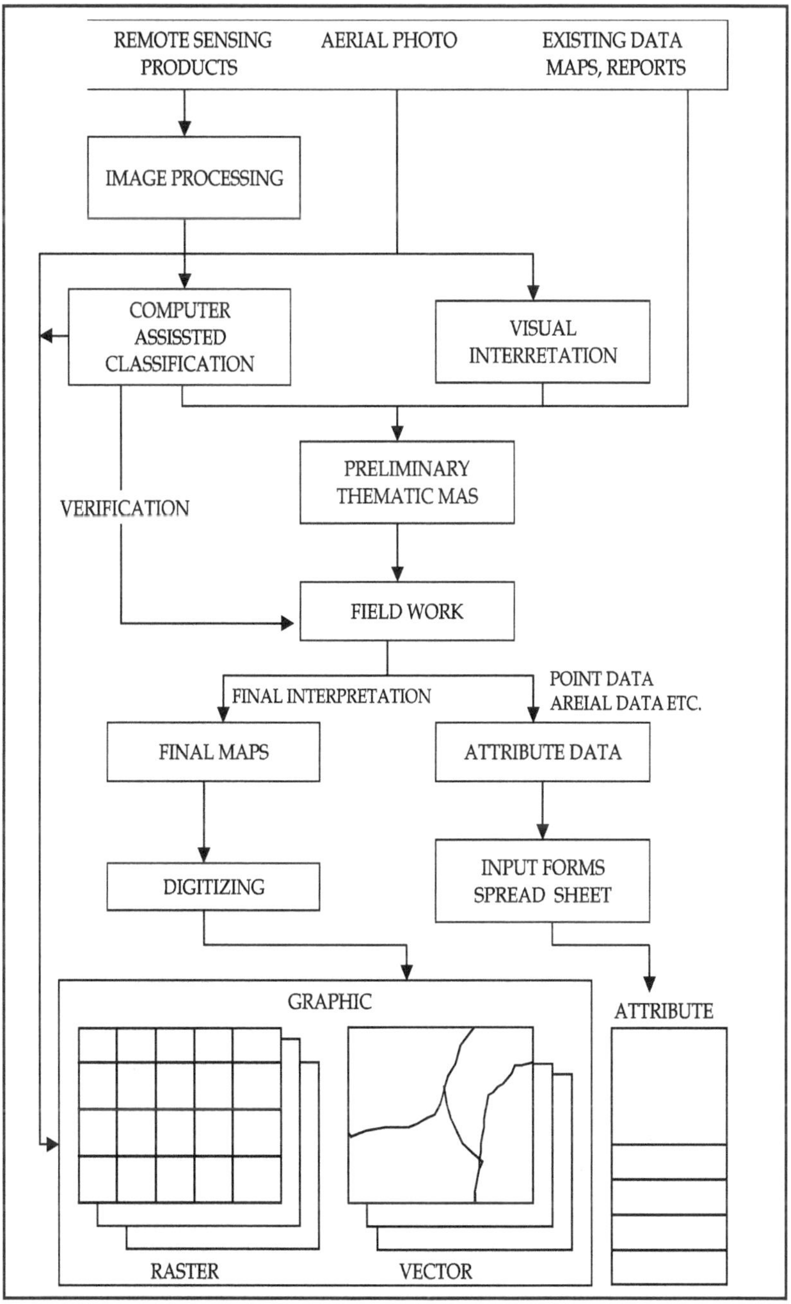

Figure 1. Remote sensing products in a GIS

command areas, alarming rate of reservoir sedimentation due to deforestation, deteriorating water quality and environment and snow cover assessment in remote and inaccessible areas. All these are required to be tackled through systematic approaches involving judicious mix of conventional methods and remote sensing data in GIS environment. The availability of remotely sensed data and use of GIS has provided significant impetus to hydrological analyses and design and their utilization in water resources planning, development and management. Remote sensing and GIS techniques have been extensively used in various areas of water resources development and management, such as land use/cover classification, flood plain management, command area studies, waterlogging and soil salinity, snow cover studies, sedimentation in reservoirs and water quality studies, etc. which are defined below in brief.

Flood management

Among all natural disasters, floods are the most frequent to be faced in India. Floods in the eastern part of India viz. Orissa, West Bengal, Bihar and Andhra Pradesh in the recent past, are striking examples. As per the information published by different government agencies the tangible and intangible losses due to floods in India are increasing at alarming rate. As reported by Central Water Commission (CWC) under Ministry of Water Resources, Government of India, the annual average area affected by floods is 7.563 million ha. This observation is based on the data for the period 1953 to 2000 published in IWRS (2001), with variability ranging from 1.26 million ha in 1965 to 17.5 million ha in 1978. On an average floods have affected about 33 million persons during 1953 to 2000. There is every possibility that this figure may increase due to rapid growth of population and increased encroachments of the flood plains for habitation, cultivation and other activities.

The main causes of floods in India are inadequate capacity within river banks to contain high flows, river bank erosion and silting of river beds. The additional factors such as land slides leading to obstruction of flow and change of the river course, retardation of flow due to tidal and back water effects, poor natural drainage in the flood prone area, cyclone and associated heavy rain storms or cloud bursts, snowmelt and glacial outbursts and dam break flows (Kumar et al., 2005).

After the disastrous floods of 1954, a national programme of flood management was launched. The government of India has taken a number of steps for flood management. As stated by Mohapatra and Singh (2003), some of the important policies on flood management include: policy statement (1954); high level committee on flood (1957); policy statement (1958); ministerial committee on flood control (1964); ministers committee on flood and flood relief (1972); working groups on flood control for five year plans; Rashtriya Barh Ayog (1980) National Water Policy (1987); National Commission for Integrated Water Resource Development (1996); Regional Task Force (1996); and National Water Policy (2002). The committees and commissions constituted by the government have given valuable recommendations on different issues of flood management. Various types of structural as well as non-structural measures have been taken up to reduce the damages in the flood plains. The structural measures, such as the construction of embankments, levees, spurs, etc. have been implemented in some of the states. The total length of constructed embankments is 16,800 km and drainage channels are of 32,500 km. A total of 1,040 towns and 4,760 villages are currently protected against flood. Barring occasional breaches in embankments, these have provided reasonable protection to an area of about 15.07 M ha. A large number of reservoirs have been constructed and these reservoirs have resulted in reduction of intensity of floods.

The non-structural measures such as flood forecasting and warning are also being adopted. The flood forecasting and flood warning in India commenced in 1958, for the Yamuna River in Delhi. It has evolved to cover most of the flood prone interstate river basins in India. The Central Water Commission has established a flood forecasting system covering 62 major rivers with more than 157 stations for issuing flood forecasts covering almost all the flood prone states. The response of state governments towards enactment of flood plain zoning bill is not encouraging. Though some of the states (e.g. Rajasthan and Manipur) have enacted the flood plain zoning legislation, but the major flood affected states (e.g. Assam, Himachal, Goa and Sikkim) have not considered such legislation. A working group of National Natural Resources Management System[15] standing committee on water resources for flood risk zoning of major flood prone rivers considering remote sensing input was constituted by the Ministry of Water Resources in 1999. The working group recommended flood risk zoning using satellite based remote sensing with a view to give thrust towards implementation of flood plain zoning measures (NNRMS, 2002).

In the conventional methods of flood risk zoning, the flood discharge is routed through the river reach to estimate the likely inundation due to spilling over the banks/embankments based on topographic contour maps and configuration of the river geometry obtained through land surveys. Continuous availability of satellite-based remote sensing data has made the understanding of dynamics of flood events much easier. The satellite remote sensing techniques provides a wide area synoptic coverage, repetitiveness and consistency, which enable the collection of information on all major flood events on a reliable basis.

Remote sensing can provide information on flood-inundated areas for different magnitudes of floods so that the extent of flooding in the flood plains can be related to the flood magnitude. Duration of flooding can be estimated with the help of multiple coverage satellite imagery of the same area within 2/3 days. High-resolution satellite data provides information on the floodplain and effectiveness of flood control works. Extent of inundation for specific flood return periods can be estimated. Using close contour information, inundation extent for given elevation, can be estimated which is a vital input for risk zone mapping. A number of studies for Ganges and Brahmputra basin have been carried out using remote sensing and GIS.

GIS provides a broad range of tools for determining area affected by floods and for forecasting areas that are likely to be

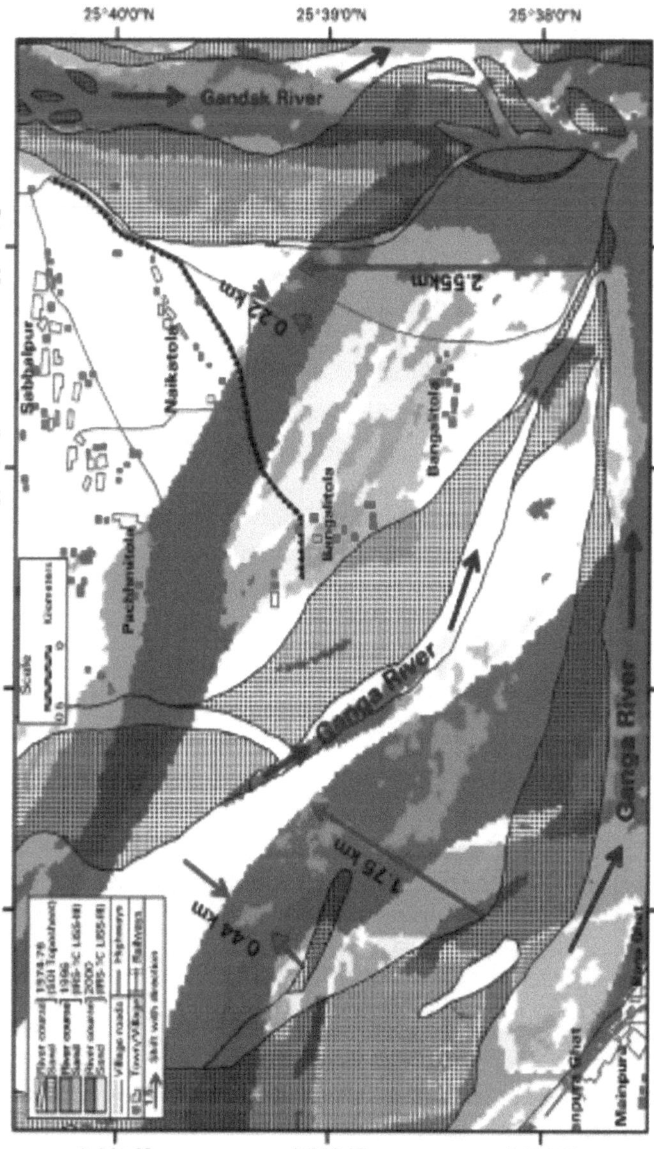

Figure 2: Details of shifting of River Ganga near its confluence with Gandak at Hazipur as obtained from Survey of India toposheets and IRS LISSIII data (1996 and 2000)

flooded due to high water level in a river. Spatial data stored in the digital database of the GIS, such as a digital elevation model (DEM), can be used to predict the future flood events. The GIS database may also contain agriculture, socio-economic, communication, population and infrastructure data. This can be used, in conjunction with the flooding data to adopt an evacuation strategy, rehabilitation planning and damage assessment in case of a critical flood situation. The NIH has completed flood inundation and river shifting studies for a number of rivers in India. The river shifting in a part of Ganga River is shown in Figure 2.

Flood management also calls for community participation. Farmers, professional bodies, industries and voluntary organizations have to be aware about flood management. People's participation in preparedness, flood fighting and disaster response is required. Media like radio, TV, newspapers can also play an important role in flood management. As India shares river systems with six neighboring countries viz. Nepal, China, Bhutan, Pakistan, Bangladesh and Myanmar; hence, bilateral cooperation for flood management is necessary for India and the concerned country. The government of India has taken some initiatives in this regard however, more active participation is required.

Drought management

The drought prone area assessed in the country is about 51.12 million ha. The planning and management of the effects of drought appear to have a low priority due to associated randomness and uncertainty in defining the start and end of droughts. Further most of the drought planning and management schemes are generally launched after persisting drought conditions. The traditional system of drought monitoring and estimating losses by crop cutting needs replacement with real time remote sensing, GIS, GPS and modeling techniques for ensuring transparency and quick response. Scope of losses may be extended to ground water depletion, damage to perennial trees, plantations, orchards and depletion in fertility of livestock. Food, fodder, agricultural inputs and water banks may be established in vulnerable areas instead of their storage in surplus regions to avoid transport bottlenecks during the drought. Robust and rainfall independent off-farm livelihood opportunities may be targeted in the drought mitigation strategy. Conjunctive use of surface and groundwater, aquifer recharge and watershed management with community participation is another important policy paradigm shift to be internalized fully.

The satellite based information, such as surface water spread and vegetation index in different years form an excellent input for monitoring drought conditions in the area. There is a need for the development of the decision support systems (DSS) for the monitoring and management of the drought on basin scale utilizing the advanced capabilities of remote sensing, geographical information system and knowledge based systems. The DSS should also provide support to the decision makers for providing the information at different spatial and temporal scales. It would help them in taking the required management measures in the drought prone areas for different administrative units. In the drought prone areas publication campaign may be launched for water conservation with the help of electronic and print media. Necessary steps may be taken at political, administrative and technical levels to encourage the people participation in the drought management for optimum utilization of the available water supply to meet the demands. Strengthening of research and development and capacity building in terms of emerging information technologies and issues of damages is also called upon to bring in resilience in the drought coping strategies.

Land use planning and cropping pattern

Another strategy that could be adopted refers to planning of land use especially in new land developments. Areas where water supply priorities are low can be planted with drought resistant varieties of trees. For this purpose drought resistant variety of trees need to be developed. Another strategy that could be suggested can be in agricultural sector. Considerable information exists on time distribution of water requirements for various crops and various planting dates. This knowledge is required to be integrated systematically with water supply probabilities to develop planting strategies. The selection of cropping pattern as per availability of water will reduce adverse impacts of drought on potential water consuming crops. The plants suitable for water scare areas can be (i) with shorter growth period (ii) high yielding plants requiring no increase in water supply (iii) plants that can tolerate saline irrigation water (iv) plants with low transpiration rates (v) plants with deep and well branched roots.

Mutli-spectral classification of land cover was one of the first well-established remote sensing application for water resources. Due to agriculture, forest management and urbanisation the physical characteristics of the land surface and upper soil as well as the evpotranspiration process, which is strongly related to the type of vegetation, are changed. As a result the amount of runoff, the soil moisture and groundwater recharge are strongly affected. Numerous investigators have used classification of land cover from satellite data as input to various water resources studies. Land use features can be identified, mapped, and studied on the basis of their spectral characteristics. Healthy green vegetation has considerably different characteristics in visible and near infrared regions of the spectrum, whereas dry bare soil has a relatively stable reflectance in both the region of the spectrum. Water shows very low reflectance in visible part of the spectrum and almost no reflectance in infrared part of the spectrum. Thus, using multi-spectral data suitably, different ground features can be differentiated from each other and thematic map depicting land use can be prepared with satellite data. Land use data devised from remote sensing are an ideal database for distributed hydrologic modelling changes. In many hydrological models the parameters of vegetation related model components (e.g. for interception and evapotranspiration) are chosen dependent on the land use. Land use/land cover information derived using remote sensing techniques has been the most widely investigated application of remote sensing in hydrologic modeling. The bulk of these efforts have centered on using the land use information to estimate SCS runoff curve numbers for stream flow forecasting.

Groundwater management

To protect the aquifers from overexploitation, an effective groundwater management policy oriented towards promotion of efficiency, equity, and sustainability is required. Agricultural holdings in India are highly fragmented and the rural population density is large. The exploitation of groundwater resources should be regulated so as not to exceed the recharging possibilities, as well as to ensure social equity. The detrimental environmental consequences of over-exploitation of groundwater need to be effectively prevented by the Central and State Governments. Over exploitation of groundwater should be avoided, especially near the coasts to prevent ingress of seawater into freshwater aquifers. Clearly, a joint management approach combining government administration with active people participation is a promising solution (Nagaraj, 1999).

In critically overexploited areas, bore-well drilling should be regulated till the water table attains the desired elevation. Artificial recharge measures need to be urgently implemented in these areas. Amongst the various recharge techniques (Oaksford, 1985) percolation tanks are least expensive in terms of initial construction costs. Many such tanks already exist but a vast majority of these structures have silted up. In such cases, cleaning of the bed of the tank will make them reusable. Promotion of participatory action in rehabilitating tanks for recharging would go a long way in augmenting groundwater supply. Due to declining water table, the cost of extraction of groundwater has been increasing over time and wells often go dry.

Remote sensing system is quite helpful in groundwater exploration as the remotely sensed data provide a large area synoptic view with high observational density. The common current remote sensing platforms record features on the surface. Most of the information for ground water, as yet, has to be obtained by qualitative reasoning and semi quantitative approaches. The remote sensing information is often of surrogate nature and has to be merged with geohydrologic data to become meaningful. The vegetation can be used as an indicator if local knowledge is available and the types can be identified on the satellite data. Apart from the contribution which remote sensing can make to understanding regional hydrogeology-necessary for managing groundwater resources-perhaps the strongest application for the management in the evaluation of the recharge, the groundwater drafts for irrigation and the identification of flow systems in areas where there is paucity of geohydrological data. Surface conditions, soils, weathered zones, geomorphology and vegetation determine the recharge, suitability for artificial recharge and soil and water conservation measures which can affect the recharge.

Groundwater vulnerability to pollution is also directly related to surface conditions. Indexing methods, group depth to water table, net recharge, topography, impact of vadose zone media and hydraulic conductivity of the aquifer (leading to the acronym DRASTIC) into relative ranking scheme that uses a combination of weights and ratings to produce numerical values.

Shah, 2005 mentions following workable solutions for management of groundwater resources:

- Banning private wells is futile; crowd them out by improving public water supply.
- Regulating final users is impossible; facilitate mediating agencies to emerge, and regulate them.
- Pricing agricultural groundwater use is infeasible; instead, use energy pricing and supply to manage agricultural groundwater draft.
- No alternative to improved supply side management; better rain-water capture and recharge, imported surface water in lieu of groundwater pumping.
- Grow the economy, take pressure off land, and formalize the water sector.

Watershed management

Proper planning of watersheds is essential for the conservation of water and land resources and their productivity. Characterisation and analysis of watersheds are a pre-requisite for this purpose. Watershed characterisation involves measurement of parameters of geological, hydro-geological, geomorphological, hydrological, soil, land cover/land use etc. Remote sensing via aerial and space borne sensors can be gainfully used for watershed characterisation and assessing management requirements and periodic monitoring. The various physiographic measurements that could be obtained from remotely sensed data are size, shape, topography, drainage pattern and landforms. The quantitative analysis of drainage networks enables relationships between different aspects of the drainage pattern of the basin to be formulated as general laws and to define certain useful properties/indices of drainage basin in numerical terms. The laws of stream numbers, stream length and stream slopes can be derived from measurements made in the drainage basin. Remote sensing along with ground based information in GIS mode can be used for broad and reconnaissance level interpretations for land capability classes, irrigation suitability classes, potential land users, responsive water harvesting areas and monitoring the effects of watershed conservation measures. Correlations for runoff and sediment yields from different watersheds versus land use changes and land degradation could also be established.

Currently, many programs, campaigns and projects are underway in different parts of India to spread mass awareness and mobilize the general population in managing water resources. Some of these are being implemented by the Central/State Governments, while others have been taken up by various Non-Governmental Organizations (NGOs). For example, Hariyali (meaning 'greenery') is a watershed management project, launched by the Government of India, which aims at enabling the rural population to conserve water for drinking, irrigation, fisheries and afforestation as well as generate employment opportunities. The project is being executed by the Gram Panchayats (village governing bodies) with people's participation; the technical support is provided by the block (sub-district) administration (Kumar et al., 2005).

Soil erosion and reservoir sedimentation

Soil is eroded due to rainfall and winds, resulting in tremendous sediment movement into water courses by flood and storm waters. A great amount of sediment is carried annually by the Indian rivers down to the reservoirs, lakes, estuaries, bays, and oceans. The impact of sediment erosion, transport and deposition is widespread. Deposition of coarse sediments reduces the reservoir storage and channel conveyance for water supply, irrigation, and navigation and cause extensive disturbance to streams. Suspended sediments reduce water clarity and sunlight penetration, thereby affecting the biotic life. As the sediment settles to the bottom of water bodies, it buries and kills vegetation and changes the ecosystem. Sedimentation reduces the storage capacity of reservoirs and hence their ability to conserve water for various intended purposes.

Conventional techniques for sedimentation quantification are: a) direct measurement of sediment deposition by hydrographic surveys, and b) indirect measurement of sediment concentration by inflow - outflow method. Both these methods are laborious, time consuming, and costly and have their own limitations. Sampling and measurement of suspended sediments is a tedious and expensive program for either in-situ or laboratory work.

With the introduction of remote sensing techniques in the recent past, it has become cheap and convenient to quantify sedimentation in a reservoir and to assess its distribution and deposition pattern. Remote sensing techniques, offering data acquisition over a long time period and broad spectral range, are superior to conventional methods for data acquisition. The advantage of satellite data over conventional sampling procedures include repetitive coverage of a given area (16/22 days) a synoptic view, which is unobtainable by conventional methods, and almost instantaneous spatial data over the areas of interest. The basic output from remote sensing data obtained is water spread area. A number of studies using remote sensing techniques have been carried out in NIH. (Jain et al., 2004). In Figure 3 image of Ghatprabha reservoir for two dates have been shown. Here we can see that the water spread area reduced from maximum to minimum in a year. On the basis of water level and water spread area, storage capacity is determined and thus the sedimentation rate.

Water conservation

Water conservation implies improving the availability of water through augmentation by means of storage of water in surface reservoirs, tanks, soil, and groundwater zone. It emphasizes the need to modify the space and time availability of water to meet the demands. This concept also highlights the need for judicious use of water. There is a great potential for better conservation and management of water resources in its various uses. On the demand side, a variety of economic, administrative and community-based measures can help conserve water. Also, it is necessary to control the growth of population since large population is putting massive stress on all natural resources.

Since agriculture accounts for about 69% of all water withdrawn, the greatest potential for conservation lies in increasing irrigation efficiencies. Just a 10% improvement in irrigation efficiency could conserve enough water to double the amount available for drinking. In India, sprinkler irrigation is being adopted in Haryana, Rajasthan, Uttar Pradesh, Karnataka, Gujarat and Maharashtra. The use of sprinkler irrigation saves about 56% of water for the winter crops of Bajra and Jawar, while for Cotton, the saving is about 30% as compared to the traditional gravity irrigation. An important supplement to conservation is to minimize the wastage of water. In urban water supply, for example, almost 30% of the water is wasted due to leakages, carelessness, etc. while most metro cities face deficit in supply of water. It is, therefore, imperative

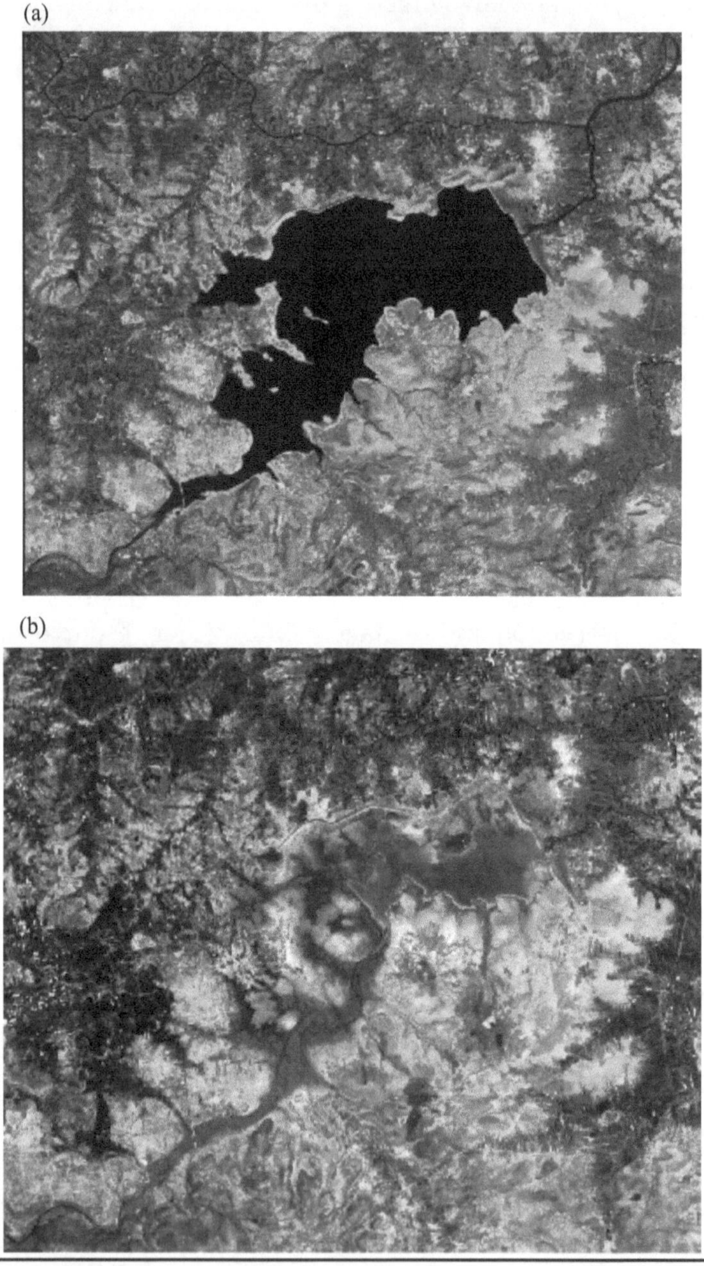

Figure 3. False Colour Composite (FCC) of Ghatprabha reservoir located in Karnataka (a) October 2000 (b) April 2001

to prevent wastage. In industries also, there is a scope for economy in the use of water. Prices of water for all uses should be fixed, keeping in mind its economic value, control of wastage, and the ability of users to pay. As water is becoming scarcer, pricing will be an important factor in avoiding wastage and ensuring optimal use.

Interbasin water transfer

The vast variation, both in space and time, in the availability of water in different regions of India has created a food-drought-flood syndrome with some areas suffering from flood damages and other areas facing acute water shortage. The States of Karnataka, Tamil Nadu, Rajasthan, Gujarat, Andhra Pradesh and Maharashtra are the worst drought prone States. The States of Uttar Pradesh, Bihar, West Bengal, Orissa and Assam face severe flood problems. Inter-basin transfer of water in India is a long-term option to partly overcome the spatial and temporal imbalance of availability and demand of water resources.

The transfer of water from surplus areas to deficit areas is not a new concept. Many such schemes have been implemented all over the world. In India too, projects like Periyar -Vaigai system, Indira Gandhi canal, and Telugu Ganga stand as classic examples of inter-basin water transfer. In the seventies, the Garland Canal proposal of Capt. Dastur and the Ganga - Cauvery Canal proposal of Rao (1973) were received with considerable attention. A National Perspective Plan (NPP) for water resources development was formulated by the Government of India in 1980s. The distinctive feature of the NPP is that the transfer of water is essentially by gravity and only in small reaches by lift pumping (not exceeding 120 m). This plan comprises of two components: (a) Himalayan Rivers Development, and (b) Peninsular Rivers Development. While the second component will be an inter-state venture, the first will involve neighboring countries too and thus will be an international venture. Some of the major benefits expected from inter linking of the rivers(Water Management Forum, 2003 and IWRS, 1996) are: (i) irrigation potential is to increase from 140 to 175 million ha; (ii) drinking water availability is to increase by about 12 km^3; (iii) peak flood discharge to get reduced by about 30% due construction of reservoirs; (iv) generation of 34000 MW of electricity; and (v) possibilities of inland navigation to provide cheap transport.

Conjunctive use of surface and groundwater

Water management in command areas requires to be given serious attention in view of the disappointing performance of irrigation projects despite huge investments. Large canal infrastructure network for providing irrigation has been the prime goal of the Government of India, since first five-year plan, which continued up to seventh five-year plan. In some of the irrigation project commands such as Sarda Sahayak in U.P., Gandak in Bihar, Chambal in Rajasthan, Nagarjuna Sagar in Andhra Pradesh, Ghatprabha and Malprabha in Karnataka etc., problems of waterlogging are being faced. The main reason for excessive use of surface water as compared to groundwater is its much lower price for irrigation as compared to the cost incurred in using groundwater. Waterlogging problems could be overcome if conjunctive use of surface and groundwater is made. Groundwater utilization for irrigation in waterlogged areas can help to lower the groundwater table and reclaim the affected soil. Over exploitation of groundwater in areas like Mehsana, in Gujarat; parts of Meeurt and Varanasi districts in Uttar Pradesh, Coimbatore in Tamil Nadu and Karnal district in Haryana etc. has resulted in mining of groundwater (Chandra, 1987). Many research workers have focused the causes of waterlogging (Raghuvanshi et al., 1990). Several groundwater flow modeling studies have focused on assessing the waterlogged areas and measures to control problems of waterlogging and salinization (Gates et al., 2002 & Dawoud et al., 2004). To asses waterlogging in command areas, multi-spectral and multi temporal remote sensing data are very useful. The satellite data thus provide a quick and more reliable delineation of the waterlogged areas and inundation. The spatial distribution of soil affected by 'positional waterlogging' (i.e., that due to its location in the landscape) can be modeled with digital topographic data using the concept of contributing area. This waterlogging depends on two topographic factors i) the local slope angle and ii) the drainage area. The probability of waterlogging increases with the contributing drainage area and decreases with increasing local slope angle. As the waterlogging phenomenon is related with topography, so Digital Terrain Modelling (DTM) can aid in detecting the waterlogged areas. DTM provide information regarding slope, aspect etc., which in turn provide information about the areas susceptible to waterlogging. In NIH waterlogging studies for IGNP, Ghatprabha command area, Muktsar, Rohatak and Jajjar districts have been carried out. The waterlogged area in Rohtak and Jhajjar districts are shown in Figure 4 (Jain *et al.*, 2004).

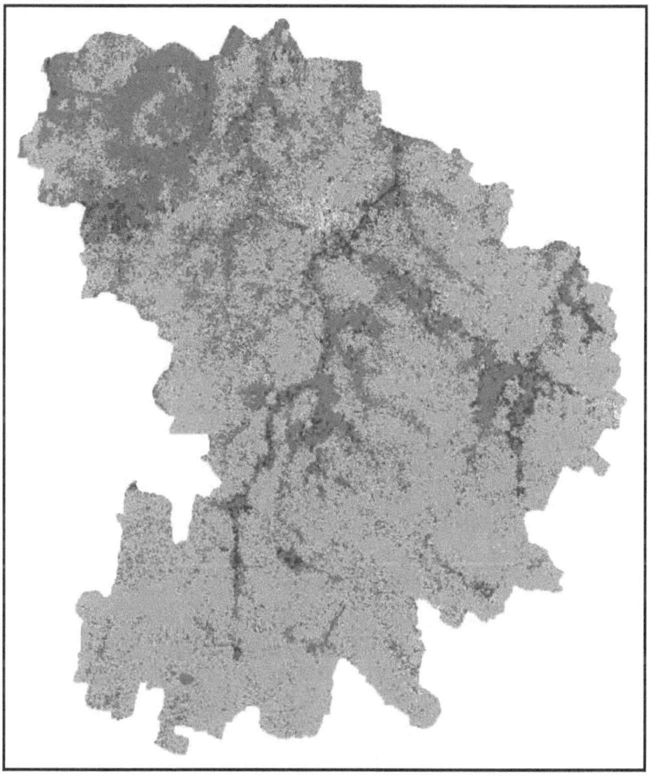

Figure 4. Waterlogged area in Rohtak and Jhajjar districts of Harayana

It is desirable that the irrigation needs for fulfilling crop water requirements should be satisfied by judicious utilization of available canal water in conjunction with ground water so as to keep the water table within the acceptable range. Thus, the optimal conjunctive use of the region's surface and groundwater resources would help in minimizing the problems of waterlogging and groundwater mining.

The command area is the total area lying between drainage boundaries, which can be irrigated by a canal system. Remote sensing can play a useful complimentary role in managing the land and water resources of command areas to maximize the production. Management of water supplies for irrigation in command areas is a critical problem to tackle with vis-a-vis limited quantities. This requires information on total demand and the distribution of demand for irrigation in command areas. Moreover the vastness of area involved, time constraints and yearly changes demand fast inventory of the situation. With more area being brought under irrigation, crop monitoring also becomes essential for estimating agricultural production and efficient planning of water management. Remote sensing can be looked upon as an aid in planning and decision-making for the aforementioned purposes. The usefulness of remote sensing techniques in inventory of irrigated lands, identification of crop types, their extent and condition and production estimation have been demonstrated in various investigations. Conjunctive use planning of surface and groundwater can be done using the remotely sensed information on surface water assessment in conjunction with ground-based data on groundwater availability. This would permit development of conjunctive use models for land water allocations in GIS environment.

GIS for Decision Making

The GIS is derived from multiple sources of data with different levels of accuracies. While a single piece of data can be assigned an accuracy value, information derived from multiple sources of inaccurate data can also be assigned an level of accuracy. In any pictorial representation of data the uncertainty can be brought in as one of the dimensions to guide the final decision-making. Any decision today has to depend on a variety of factors, which are available in an information system like GIS.

Recently, new advances in computer technology have enabled widespread improvement in water resources planning and management. One of the new trends in solution of water management problems has been to aggregate several models into integrated software i.e. a knowledge based Decision Support System (DSS) that focuses on the interaction between the use and the data, models, and computers. Rapidly advancing computational ability, development of user-friendly software and operating systems, and increased access to and familiarity with computers among decision makers are the important reasons for this growth in the fields of both research and practice. Automating the process with a DSS could effectively improve the water resources planning and management.

The aim has been to create a system in which the mechanics of linking one component with another are largely transparent to the user. Although time and effort is required in customizing the system for a particular river basin, its use has been designed to be as simple as possible. Communication is by means of a user-friendly interface, which makes extensive use of hypertext for guidance and colour graphics in presenting the results. A keyboard is not normally required since all the facilities are accessed by touching the appropriate icon representing the particular component. Moreover, the controls for operating a model will be familiar in as much that the icons mimic a video-recorder viz. play, pause and stop. In this way, the person responsible for the actual project is able to make rational use of the system without an in-depth knowledge of modeling techniques.

Comments

In view of the existing status of water resources and increasing demands of water for meeting the requirements of the rapidly growing population of the country as well as the problems that are likely to arise in future, a holistic, well-planned long-term strategy is needed for sustainable water resources management in India. The water resources management practices should be based on increasing the water supply and managing the water demand under the stressed water availability conditions. Data monitoring, processing, storage, retrieval and dissemination constitute the very important aspects of the water resources management. These data may be utilized not only for management but also for the planning and design of the water resources structures. In addition to these, now a days decision support systems are being developed for providing the necessary inputs to the decision makers for water resources management. Also, knowledge sharing, people's participation, mass communication & capacity building are essential for effective water resources management.

Integrated and coordinated development of surface water and ground water resources and their conjunctive use should be envisaged right from the project planning stage and should form an integral part of the project implementation. There should be a periodic reassessment of the surface and ground water potential on a scientific basis, taking into consideration the quality of water available and economic viability of its extraction. Since the hydrological processes are continuous and quite complex, therefore, an accurate assessment of quantities of water simultaneously passing through all these processes is quite a difficult task. Mathematical modelling of hydrological processes would provide an opportunity to both the research hydrologists and the water resources engineers involved in developing the integrated approaches for planning, development and management of water resources projects for sustainable development as well for preserving the eco-systems.

A combination of remote sensing data, ground truth, (e.g. from modern hydrometric equipment) and powerful software domains such as GIS give the water resources manager extremely powerful tools to help solve their problems. The reason of adopting GIS technology is because it allows the spatial information to be displayed in integrative ways that are readily comprehensible and visual. Remote sensing is now being widely regarded as a layer in the GIS. Although remote sensing is a specialized technique, it is now being accepted as a basic survey methodology and as a means of providing data for a resource database. Decision Support Systems are required to be developed for planning, development and management of the water resources projects. (Tsihrintzix *et al.,* 1996)

The research needs in the area of water resources, involving remote sensing and GIS inputs, are as follows:

- First and very important aspect is the data availability and compatibility in any GIS related study. Spatial information required for water resources studies should be readily available for timely execution. Remote sensing can provide many of the necessary data to supplement conventional data and also provide new data types and forms that will help to tackle previously unsolvable questions.
- The future progress in the hydrological sciences will depend upon the availability of adequate data for model development and validation. Remote sensing can and should play a pivotal role in this progress. The data banks should provide digitized maps and their spatial data compatible with various systems. Such data availability could significantly speed up the analysis.
- One difficult task in incorporating GIS in water resources modelling is the interfacing of water system models with the GIS. Automation of interfacing tasks is one of the areas to be researched in incorporating GIS and available models. The recent developments of Decision Support Systems (DSS) to assist water resources decision making hold the key for integration of GIS and water resources models. Such systems provide an integrated approach for water resources management considering the various water related disciplines together with socio-economic aspects. These systems may be utilized for studying the different scenarios of water demand for arriving at the optimum allocation of water for various demands under varying water availability conditions.

- Another area of potential research to further enhance the modelling process is the integration of expert systems and GIS. Expert GIS systems can be used to provide regulatory information by linking regulatory facts stored in a database to sites located in a GIS through an expert system query interface.
- Further research is needed for comparing the GIS packages available in the market and their characteristics, providing check lists for GIS users.

References

CGWB, Ground Water Resources of India, Central Ground Water Board, Govt. of India, New Delhi, 1995.

Chandra, S. Planning for integrated water resources development project with special reference to conjunctive use of surface and groundwater resources, Central Ground Water Board, New Delhi, 1987.

Dawoud, M.A. and Allam, A.R., Effect of new Nag Hammadi Barrage on groundwater and drainage conditions and suggestion of mitigation measures, *Wat. Resour. Manage.*, 18(4), 321-337, 2004.

Gates, T.K., Burkhalter, J.P., Labadie, J.W., Valliant, J.C. and Broner, I., Monitoring and modeling flow and salt transport in a salinity threatened irrigated valley, *J. Irrig. Drain. Engng.* ASCE, 128(2), 87-99, 2002.

IWRS, Theme Paper "Five Decades of Water Resources Development in India". Indian Water Resources Society, Roorkee, 1998.

IWRS, Theme paper on Management of floods and droughts, Indian Water Resources Society, Roorkee, India, pp. 4-12, 2001.

IWRS, Theme paper on "Inter Basin Transfers of Water for National Development - Problems and Prospects". Indian Water Resources Society, Roorkee, 1996.

Jain, S.K., M. K. Goel, V. S. Jeyakanthan, T. R. Nayak and Dilip Durbude, 2004, Assessment of sedimentation in Ghatprabha, Gandhi sagar, Vaigai, Tandual and Lingnamakki reservoirs using digital image processing techniques, Report-CS/AR-1/2003-04, National Institute of Hydrology, Roorkee.

Jain, S. K., D. S. Rathore and K. D. Sharma, 2004, Remote sensing and GIS applications in the management of water resources development, *ISG Newsletter*, 10, 1 & 2, March/June 2004

Kumar Rakesh, R. D. Singh and K. D. Sharma, 2005, Water resources of India, *Current Science*, A fortnightly multidisciplinary journal (in press).

Lal, M., Climate Change - Implications for India's Water Resources. *J. of India Water Resources Society*, 2001, 21, 3, July, pp. 101-119.

Mohapatra, P.K. and Singh, R.D., Flood management in India, J. Natural Hazards, Kluwer Academic Publishers, the Netherlands, 28, 131-143, 2003.

National Water Policy, Ministry of Water Resources, Government of India, 2002.

NNRMS, Flood risk zoning using satellite based remote sensing, Ministry of Water Resources, Govt. of India, 2002.

Nagaraj, N., Marshall Frasier, W., and Sampath, R. K. (1999) A Comparative Study of Groundwater Institutions in the Western United States and Peninsular India for Sustainable and Equitable Resource Use. Department of Agricultural and Resource Economics, Colorado State University, Fort Collins, Colorado, USA.

NCIWRD, Integrated Water Resources Development - A Plan for Action. Report of The National Commission for Integrated Water Resources Development. Govt. of India, Ministry of Water Resources, New Delhi, 1999.

Oaksford, E.T., Artificial Recharge: Methods, Hydraulics, and Monitoring, In: Artificial Recharge of Ground water, T. Asamo, editor. Butterworth Publishers, Boston, 1985.

Rashtriya Barh Ayog, Report Vol. 1. Ministry of Energy and Irrigation, 1980.

Raghuvanshi, C.S., M. Chandra and Raghuvanshi, T.K., Socio-economic impact of waterlogging - A critique of mismanagement of surface irrigation. Proceedings of All India Seminar on Waterlogging and Drainage, Roorkee, 1990.

Rao, K L., India's Water Wealth. Orient Longman Limited, New Delhi, 1973.

Rango A., 1990, Remote sensing of water resources: Accomplishments, challenges and relevance to global monitoring, Proceedings

International symposium remote sensing and water resources, Enschede, The Netherlands, August 20-24, 1990.

Schultz G.A. and E.T.Engment (EDS.), 2000, Remote sensing in hydrology and water management, Springer Verlag.

Sinha, A.K., V.M. Ghare, D. Dhar, V.D.Khare and S. A. Patil, 2004, Remote Sensing and GIS for water resources management, ISG Newsletter, Indian Society of Geomatics, Vol. 10, No. 1 & 2, March/June 2004.

Shah, T., Groundwater and human development: Challenges and opportunities in livelihoods creation and environment. Lecture delivered during the Workshop on Creating Synergy between Groundwater Research and Management in South Asia, National Institute of Hydrology, Roorkee, India, 2005, 7-9 February.

Seth, S.M., Integrated water resources management - role of research and development in hydrology. Proceedings of International Conference on Integrated Water Resources Management for Sustainable Development, New Delhi. Organized by National Institute of Hydrology, Roorkee, 2000.

Tsihrintzix, V.A., R. Hamid and H.R.Fuentes, 1996, Use of GIS in water resources: A review., Water Resources Management 10: 251-277, 1996.

Water Management Forum, Inter-basin transfer of water in India - Prospects & Problems". New Delhi, February, Water Management Forum. The Institution of Engineers (India), 2003.

Glossary

Geographic Information System: An organized collection of computer hardware, software, geographic data, and personnely designed to efficiently capture, store, update, manipulate, analyze, and display all forms of geographically referenced information.

Digital Elevation Model (DEM): Representation of the topography of the earth or another surface in digital format, i.e. by coordinates and numerical descriptions of altitude. DEMs are used often in GIS. A DEM may or may not be accompanied by information about the ground cover. In contrast withtopographical maps, the information is stored in a raster format. That is, the map will normally divide the area into rectangular pixels and store the elevation of each pixel. In that sense, digital elevation model (DEM) data are sampled arrays of surface elevations in raster form.

Remote Sensing: Measurement or acquisition of information of an object or phenomenon, by a recording device that is not in physical or intimate contact with the object.

Digital Terrain Model: Digital representation of a topographic surface. It is commonly used interchangeably with DEM. Strictly speaking a DIM refers to a model of reality, which includes information relating to factors such as surface texture as well as elevation.

Global Positioning System: A worldwide radio-navigation system that was developed by the U.S. Department of Defense. In addition to military purposes it is widely used in marine, terrestrial navigation and location based services

Conjunctive Use: The combined use of surface and ground water systems to optimize resource use and minimize adverse effects of using a single source.

Decision Support System: A computer-based system that aids the process of decision making, it as an interactive, flexible, and adaptable computer-based information system, especially developed for supporting the solution of a non-structured management problem for improved decision making. It utilizes data, provides an easy-to-use interface, and allows for the decision maker's own insights.

❑❑❑

GEOINFORMATICS APPLICATIONS IN AGRICULTURE, 2007
Eds. : A.K. Singh and U.K. Chopra; pp. 209-243
New India Publishing Agency, New Delhi (India)
E-mail : newindiapublishingagency@gmail.com
Web: www.bookfactoryindia.com

8

Application of Geospatial Tools, Techniques and Concepts for Integrated Watershed Management

A. Sarangi & A.K. Singh

Key words : Integrated Watershed Management (IWM), GIS, DTM, Watershed Delination, Water Quality Modeling, RS, DSS, Ground Water Information System (GWIS).

Introduction

The availability of land and water resources has a major effect on the economic and ecological aspects of life everywhere around the world. As a result, the need to develop better tools for efficient land and water resources management has become very vital. For quicker, reliable and precise results, it is necessary to use the geospatial tools and techniques such as Global Positioning Systems (GPS), Geographic Information Systems (GIS) and Remote Sensing (RS) image interpretation and analysis techniques to manage the natural resources. The application of these tools and techniques not only facilitates timely and repetitive analysis of all the topological features, but also assists in development of Decision Support Systems (DSS) to generate alternative scenarios for optimal use of the natural resources in the long term. The optimal use of these resources through integrated management of land water and vegetation leads to enhanced production and productivity of food, fodder, fiber, fuel, fish, flower and forest resources.

Moreover, the holistic management of the resources to produce these outputs is plausible and manageable within the watershed systems, which are hydro geological units from which the runoff resulting from precipitation flows past through a single outlet to downstream locations. The systematic management of land, water, vegetation and livestock resources within the watershed systems is termed as Integrated Watershed Management (IWM). Broadly, IWM can be defined as a multidisciplinary, holistic way of protecting and managing a watershed's natural resources to enhance biomass production in an eco-friendly manner. The watershed is viewed as a hydro-geological complex and dynamic ecosystem in which natural and anthropogenic processes occur and interact resulting in generation of runoff at the watershed outlet. The watershed also acts as a social, economic and political unit for planning and management purposes. Hence, the watershed manager and policy makers must consider all technical, social, economic, environmental, legal and institutional aspects of watershed planning processes (Sarangi *et al.,* 2004a). Concern about widespread soil degradation and scarce, poorly managed water resources has led to the implementation of watershed management investments throughout Asia, Africa and Latin America (Kerr and Chung, 2001). Despite the growing importance of watershed projects as an approach to rural development and natural resource management, until now there has been relatively little research on their impact (Kerr and Chung, 2001). On the whole, an IWM system will (i) control sediment outflow to prevent downstream flooding, (ii) conserve water and soil fertility, (iii) facilitate crop management and use land effectively, (iv) assist in soil and water resources planning and management, (v) reduce environmental and ecological degradation, and (vi) resolve issues related to stakeholder conflicts.

In the Indian context, the per capita availability of renewable freshwater has fallen drastically over the last 50 years. The water table is falling rapidly with unregulated over-exploitation of groundwater in many blocks of the country. India receives about 400 million hectare-metres (Mham) of precipitation annually, which is augmented by 20 Mham contributed by rivers flowing in from the neighbouring countries. Net evapo-transpiration losses are nearly 200 Mham. About 135 Mham is available on the surface and the remaining recharges groundwater. It is also estimated that despite constructing a large irrigation network comprising of reservoirs, barrages and small water storage structures, close to 92 Mham of the available surface water ultimately goes to the seas. Besides these estimates, the problems with land and water availability are becoming more stressed now with population touching 1.2 billion and the situation will worsen by 2050 (projected population of1.6 billion). Due to this demographic structure of India, the per capita water and land availability will decline to water scarcity levels (per capita water availability less than 1000 m^3 per year) and unsustainable land holding level (less than 0.7 ha of land per person per year) by the year 2050. Therefore, to find a feasible solution to these problems, besides the biotechnological and other advanced crop science based approaches, there is an urgent need to bring the wastelands, watersheds and wetlands under active production process through management of land and water resources. Such development will ensure food grain availability for the present generation without leaving any major stress on the future generations to meet their food and livelihood requirements. Therefore, IWM is the right approach for sustainable development, because the watershed is the hydro-geological unit that harbors these natural resources and is therefore attracting attention, necessitating a cross-disciplinary approach to data collection and analysis (Wilson and Droste, 2000). IWM is generally recognized as the most practical and efficient way to improve water quality and other environmental indicators while maintaining regional economic viability. This approach integrates information and knowledge across several disciplines and spatial scales to understand biophysical processes, environmental impacts of alternative management systems and behavioral response of landowners to policy changes. IWM is not an exact science but an approach to environmental management that uses the topographically delineated area drained by a stream system as both the physical and analytical boundary of analysis. It encourages examination of all biophysical and socio-economic linkages such as those that exist among natural resource sectors (e.g. forestry, fisheries, agriculture and water supply), or those that exist between upstream activities and downstream impacts. The elements of integrated watershed management are classified under following classes:

A trans-disciplinary focus: meeting the challenge of IWM requires collaboration among specialists in widely varying disciplines including biological sciences, engineering, geoscience's, economics, sociology, law and ethics.

The balancing of social, economic and environmental values: as IWM processes strive to balance the full spectrum of social, economic and environmental considerations, the necessity to make trade-offs becomes inevitable and new approaches are required.

An emphasis on strategic action: as government and non-government agencies are forced to manage with ever decreasing financial and human resources, IWM processes must find ways to target key issues and tasks that are essential to success.

The need to cope with uncertainty and complexity: water resource managers are increasingly acknowledging that understanding of complex ecosystem and socio-economic systems are riddled with uncertainty, and that new analytical tools and approaches are required.

An ecosystem management approach: IWM stresses the need to consider *all* natural resources within the watershed, and necessitates that environmental considerations (e.g., the maintenance of biodiversity and ecological function) be treated equally and simultaneously with human considerations (e.g., economic and social development).

A need for more devolved decision making and consensual processes: the use of consensual approaches to planning and decision making is clearly set as IWM processes are increasingly challenged to address natural resource development and community development issues simultaneously.

Innovative institutional structures and processes that are more resilient: for any of the above to be achieved there is prerequisite for new institutional structures and processes to emerge that are responsive to localized requirements, yet consistent across broader jurisdictional reaches, dynamic and flexible, yet focused and effective.

It is understood that the IWM concepts can be implemented and possible scenarios of better management can be generated when the data of the rainfall, runoff, soil loss, pollutant load from surface and in ground water, aerial extent of the land use/land cover and soil types are available for analysis. The data generation for management of these natural resources can be accomplished by field surveys, installation of hydrologic gauging stations and pizeometers for surface and ground water quantity and quality information and through personal interaction with the watershed beneficiaries. However, the topological features of the surface, land use and land cover information, soil types and ground water table depths can be acquired from remote sensing image analysis using the ground truth data of the surveyed area. GIS, GPS and RS software and hardware tools in co-ordination with the ground truth information can be used to replace the extensive and exhaustive field data collection by manual surveys. For many locations, manual survey is not possible for inaccessible areas with adverse land terrains. These geospatial tools and techniques along with selective ground truth information will be able to replace the detailed and extensive manual survey to develop a digital data base of watershed. Further, the acquired digital data of topological features, land use/land cover changes and soil types along with measured rainfall, runoff, sediment yield and pollutant information can be used to predict the future watershed responses through use of process based hydrologic prediction models. Finally, all these measured and simulated data sets of the watershed system can be clubbed together within a decision support system framework to generate alternate scenarios of IWM leading to optimal use of land and water resources within the watershed systems. This also assists in locating the appropriate soil and water conservation structures and best management practices (BMPs) to efficiently use the soil and water resources in a watershed. Realizing the importance of watershed development programmer, this chapter deals with the applicability of these geospatial tools (GIS, RS and GPS) along with DSS (known as 4S's because the letter "S" is contained in all these four geospatial tools) leading to integrated water resources development and management of watershed systems.

Global Positioning Systems (GPS) in IWM

Global positioning systems (GPS) are the tools to locate the point on the ground with reference to the satellite locations in space. The location is expressed in latitude and longitude coordinates. This is a constellation of minimum twenty four (generally there are three to four spare satellites making a total of 27 or 28) satellites orbiting the earth at a height of 12,600 miles; five monitoring stations (such as Hawaii, Ascension Island, Diego Garcia, Kwajalein and Colorado Springs); and individual receivers. The GPS works by reading the radio signal broadcast from as few as three of these satellites simultaneously through a receiver on earth, which can pinpoint its exact location on the ground. However, the person using the GPS to locate the points on earth need not even think about the satellites or the monitoring stations. Just like radio signals, the GPS signals are available to anyone who has the proper equipment and knowledge.

GPS works on a simpler principle in which the length of the travel time for the signal to reach the receiver from the satellite used to calculate the satellite's distance from the receiver. GPS satellites transmit continuous streams of coded radio signals that indicate their position in space and the exact time that signals are being sent.

Theoretically, with a perfect clock, signals from three satellites would be enough information for a receiver to compute its position and express it as a co-ordinate of latitude and longitude. But, the perfect clock concept does not quite exist and a discrepancy between satellite and receiver timing of just $1/100^{th}$ of a second could make a misreading of 1860 miles. Such measurement errors necessitate the fourth satellite which could fix an elevation. Moreover, the errors due to not-quite-constant speed of light, faulty clocks, ephemeris variations, receiver fuzziness and multi-path "ghosting" can be rectified by using a differential GPS (DGPS), in which the differential corrections are carried out through post processing or real time processing (Fig. 1). In real time processing, the differential corrections are carried out during the data collection and in post processing the corrections are made after the data has already been collected. Real time correction is mainly for applications like navigation, where while moving one can locate the position. Post processing is for applications related to mapping or surveying and is more accurate than the real time processing. The accuracy of post processing is due to the reason that, the equipment remains in that position for more time in comparison to the real rime processing when the object is on motion (e.g. navigation). With increase in recording time over the position, more satellite differential readings are available to average when the data is post processed, resulting in more accurate location of the object.

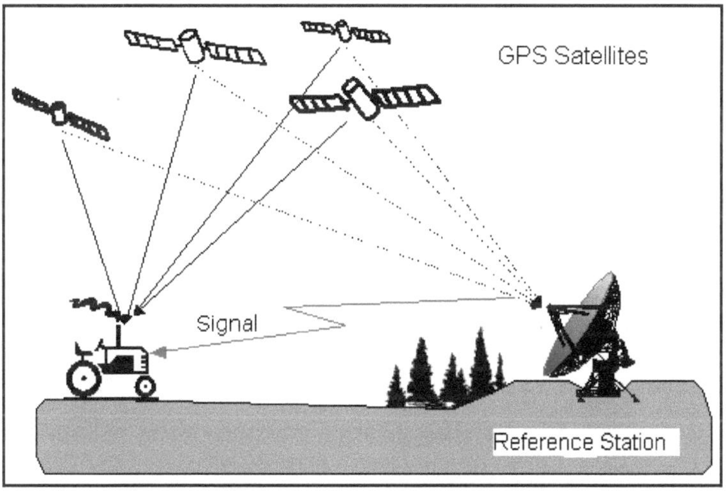

Figure 1. The schematic representation of a DGPS system

In areas where the real time differential correction service is not available, the post processing is carried out. It is observed that the post processing generated point information is more accurate than the real time processing. Though the GPS satellites are sun synchronous, the geo-stationary satellite signals are also used by third party sources for differential corrections. Moreover, the accuracy of the point co-ordinates measured by the GPS depends upon the type and quality of the GPS receiver hardware. The GPS can give the elevation of the point with reference to a known datum and also the elevation differences, but the horizontal accuracy is one to three times better than the vertical accuracy. This is because there are no satellites that we can access below the visible horizon (the earth itself blocks them), limiting the ability of the GPS receiver for accurate height measurements. The data acquired by the GPS are ultimately transferred to the computing systems through the compatible software for further processing.

GPS acquired data analysis using GIS

It is understood that the GPS receivers are used to locate the latitude and longitude values of the ground points more accurately and the point elevations less accurately. This capability of GPS assists the researcher in locating the points on map with the latitude-longitude values of the ground surface and projecting the digitized map to real world co-ordinate systems for acquiring the actual length of line, area of polygon and the distance of points on earth surface. The point location identified through the GPS can give rise to the secondary data features of that location (*e.g* soil moisture content, soil type, irrigation water depth, surface runoff depth, ground water table depth, water quality, location of conservation structures etc.) for subsequent analysis. These point features can easily be processed using advanced geostatistical techniques to predict the spatial and temporal variability on a watershed scale. Also, the GPS assisted elevation values of the watershed topology can be used to plot contours of desired interval. The contours thus generated can be used to present the watershed topology through generation of the digital elevation model (DEM). The DEM acts as a pre-processor of different hydrologic and water quality models for holistic watershed management (Garg et al, 2003; Borah and Bera, 2004). Also, the 3-D surface generated through the DEM analysis assists in deciding the location of different soil and water conservation structures to arrest sediment loss and conserve water at appropriate locations. The GPS acquired points are also used for matching the surface features for RS image analysis as a ground truth and also locating the feature classes in GIS based analysis. The watershed management requires information of the soil, water and vegetation resources extending over the watersheds for chalking out best management practices for optimal utilization of these resources for enhanced production. The acquisition of such information can be efficiently carried out by the GPS receivers. The data points or combination of a set of points provide adequate information about what is at a point, what comprises a line and polygon on the land surface in terms of soil type, land use and water quality and quantity. The spatial variability map of soil phosphorous saturation (P_{sat}) % over the St. Esprit watershed located in Quebec, Canada was generated using the differential GPS (DGPS) acquired P_{sat} values representing small land parcel of 0.4 ha (Fig. 2). The data of both latitude

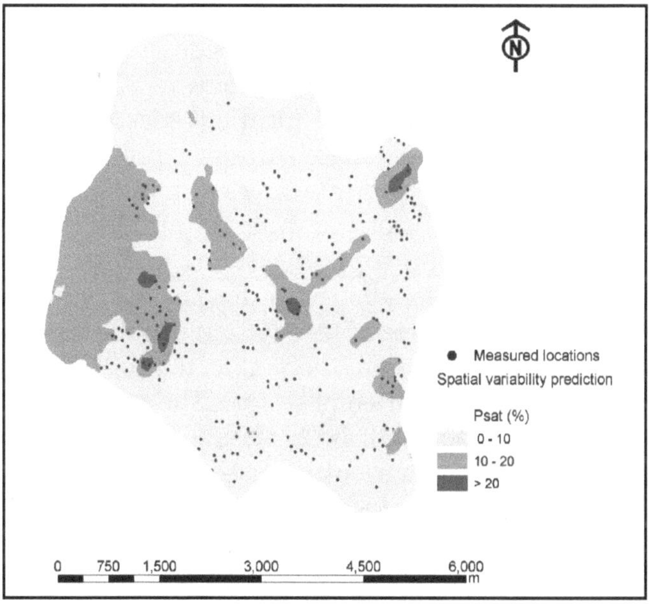

Figure 2. The Phosphorous variability map generated through GPS acquired points of the St. Esprit Watershed, Quebec, Canada.

and longitude acquired through DGPS along with the P_{sat} percentage of soil measured in laboratory as a representative of the soil sample grids for 276 locations over the watershed were used for generation of this spatial variability map using ArcGIS® (ver. 8.3) (*www.esri.com*). The P_{sat} threshold values were identified over the variability map for effective P management in the watershed. The P_{sat} values below 10% are considered to be the safe range and ranges between 10% to 20% are called supra-optimal level and the values of P_{sat} beyond 20% is considered to be critical for agro environment, as it causes eutrophication problems in the down stream water bodies of the watershed.

Geographic Information System (GIS)

A GIS consists of a package of computer programmes with graphical user interfaces (GUI) or command languages, consisting of programming codes for operating the maps, digital images, tables and geo-coded data items that provide access to particular functions. The ability of the GIS to store, analyze and integrate spatial and temporal data and generate meaningful conclusions has gained wide acceptance by the research community for management of watersheds. However, the versatility of the GIS to perform advanced data analysis through linkages with real time hydrologic prediction models, development of dynamic interfaces for integrated water resources management, optimal water resources allocation and multi-criteria evaluation support systems, and use of built in macro programming languages to develop decision support systems are under progress. Since 1982, when the GIS concepts first entered research arena, the application of GIS has been successfully applied to different sectors. The application of GIS for watershed management is considered to be an important application under the agricultural sector for enhancing the biomass production and productivity, because the management of multi-land use watersheds typically involves the co-evolution of socio-economic-ecological systems that operate at multiple spatial-temporal scales. These studies are plausible within GIS, due to its great potential for storing huge multi-sourced data in more integrated way and extend the analytical scope for different management problems at multi scale levels and subsequent integration. Also, recent advances in GIS in linking simulation models, remote sensing data, and relational database techniques make it easier for the analysis. It is fully conceivable that, within a properly integrated GIS framework, the sustainability profiles of multiple land use watersheds can be assembled into various scenarios, which may enable the basin managers to identify and avoid possible losses of significant ecological functions while defining management responses and boundaries (Diplas, 2002).

GIS in watershed delineation and hydrologic data acquisition

As explained earlier, the watersheds are defined by natural hydrological units from which the runoff water resulting from precipitation events flows past through a single outlet. It is essential to delineate the watersheds within drainage basins or big catchments as they represent the most logical basis for managing water and other natural resources. Watershed delineation can be carried out from digitized topological contour maps or Digital Elevation Models (DEM), which are defined as "*any digital representation of the continuous variation of relief or different attribute source data over space*". However, the variation of the surface terrain over space is termed as Digital Terrain Model (DTM). The DEMs can be easily generated from the digitized contour maps using the GIS tool or can be generated from remote sensing data acquired through stereo photogrammetry techniques or from aerial photographs. Several concepts exist for watershed delineation and drainage network determination from DEM or digitized contour maps (Lea, 1992; Martz and Garbrechet, 1992; Sarangi *et al.*, 2004b). The D-8 method (where the water flows in eight possible directions) is described as a single flow direction method where each cell discharges or spills onto one of its eight neighbours, the one located in the direction of steepest descent. The flow directions and generation of drainage network using the D-8 method is presented in Figure 3 with cell codes of a representative DEM.

The Rho8 method is a single flow direction method similar to D-8 method, with the difference that it determines the receiving cell among the neighboring cells based on a probabilistic approach. The Lea method (1992) is also a single flow direction method and routes the flow according to local aspect. There are also a number of multiple flow direction methods, in which the flow is distributed from a cell to its lower elevation neighbors according to some specified rule relating the cell morphology. However, the D-8 method is widely used in distributed watershed modeling as this method provided similar results in regions of convergent flow and along well defined valleys. Moreover, the developed interfaces with ArcGIS, such as WSDT (Watershed and Stream Delineation Tool), WMET (Watershed Morphology Estimation Tool) and TauDEM (Terrain analysis using Digital Elevation Model) also use D-8 and D-infinity methods for watershed demarcation and stream network generation and written in Visual Basic for Applications (VBA) built-in macro available within ArcGIS® tool. Further, the Facet

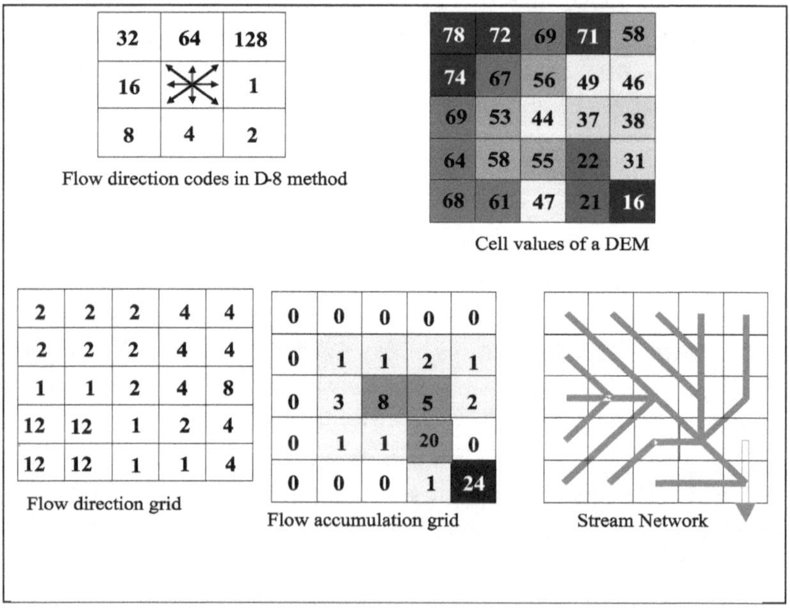

Figure 3. Watershed Delineation concepts of flow direction and flow accumulation explained for a DEM using D-8 method.

watershed data triangle method (*www.facet.com*) is a recent technology for watershed delineation and drainage network estimation which incorporate a model which considers pouring of "virtual water" over the "virtual land" and records where it flows. The terms "virtual land" and "virtual water" refers to the imaginary land area and the water amount, which can be contained within the topological surfaces of a TIN (Triangulated Irregular Network) grid. The entire watershed is converted to TIN grids of desired size. It is assumed that the virtual water drops on the triangles and depending on the shape and slope of the triangles, water trickles downhill and the flow path is recorded. Also, when the drop crosses into a new triangle, its path takes a new direction. This sequence was followed until the drop hit a water body to generate the drainage network over the surface terrain. A delineated watershed along with natural drainage network from a DEM is shown in Fig. 4 and the delineated St. Esprit watershed with Strahler's stream network is shown in Fig. 5. Using this information, the watershed morphological parameters are estimated for Banha and St Esprit watershed and presented in Table 1 (Sarangi *et al.*, 2005). These parameters are further associated with the rainfall depth, duration and intensities by empirical, conceptual, process based models and Artificial Neural Network (ANN) tools to estimate the runoff and sediment losses from watersheds. Also, the geomorphological parameters give an insight into the hydrological responses of watershed leading to watershed prioritization. For example, referring Table 1, the mean slope of St. Esprit watershed is less than that of Banha watershed, indicating the former to be a flatter watershed than the latter one. Also, the circulatory ratio of St. Esprit watershed is more than the Banha watershed. These parameters indicate that the Banha watershed is elongated watershed with higher slope and will result in quicker translation of runoff and higher peak flows for a given rainfall event. The higher hypsometric integral value of Banha watershed indicate that the watershed is more erosion prone than the St. Esprit watershed. Similarly, the qualitative information of hydrologic responses such as runoff volume, peak flow and runoff yield and sediment yield

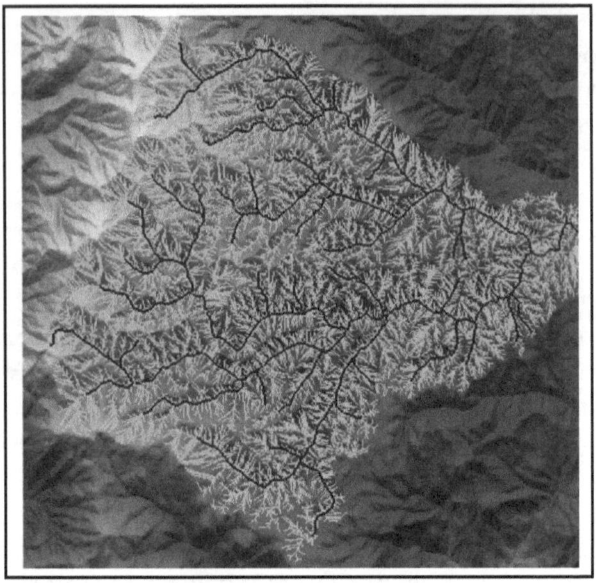

Figure 4. Watershed drainage network generation using the facet watershed triangle technique from a DEM.

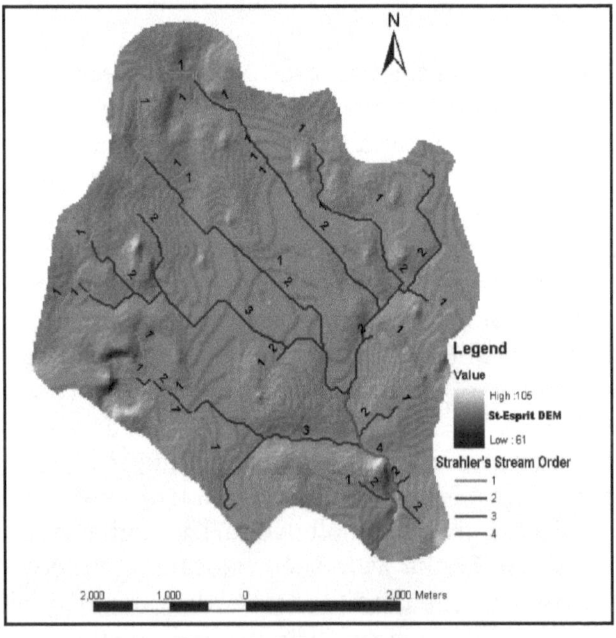

Figure 5. The drainage and stream order network of the St. Esprit Watershed, Quebec, Canada.

can be obtained by the interpretation of the estimated geomorphological parameters.

These geomorphological parameters are also used for locating the appropriate soil and water conservation measures (both mechanical and vegetative) to arrest soil loss and conserve water at appropriate locations within the watershed.

GIS in hydrologic and water quality modelling

Models are the abstracts of the real world phenomenon and an accurate mathematical model representing the watershed hydrologic processes can act as an alternative to continuous recording of real field data, which in turn can minimize the cost, time and effort involved in continuous hydrologic monitoring. A plethora of models are available for quantifying the hydrologic responses of watersheds (Haan, 1995; Borah and Bera, 2004), but only a couple of models have been developed with GIS interfaces for estimation of these parameters within a single environment. The input data required for operation of hydrologic/water quality (H/WQ) prediction models are land topology, soil type and land use information besides the climatic data. In order to facilitate the model operation and prediction, the H/WQ models are now increasingly used along with the Geographical Information System (GIS), to assist the analysis of data, test scientific hypothesis, improve the understanding of environmental processes and evaluate best management scenarios (Garg *et al.*, 2003; Sole and Valanzano, 1996). It is also understood that the hydrological behaviour of watersheds are related to interaction of the precipitation with the land topology and surface conditions. Therefore the GIS software through a set of methods or procedures is becoming the first choice of the researchers for watershed hydrological response estimation. The popular softwares available with GIS interface to perform watershed hydrological predictions are Soil and Water Assessment Tool (SWAT) (*www.brc.tamus.edu/swat*), AGricultural Non Point Source Pollution (AGNPS) model (*www.sedlab.olemiss.edu/AGNPS.html*), Spatial Water Budgeting

Model (SWBM) (http://gisweb.ciat.cgiar.org/cross_scale/download), Watershed Modelling Systems (WMS) (*www.scientificsoftwaregroup.com*) etc and a couple of models are in process of development.

Table 1. Geomorphologic parameters estimated using the WMET interface (ver 2.0) for St-Esprit and Banha watersheds.

Geomorphological Parameters	Banha Watershed, Jharkhand, India	St-Esprit Watershed, Quebec, Canada
Area (Sq. km.)	17.51	26.093
Perimeter (km)	20.05	23.679
Maximum length (km)	7.04	7.35
Maximum elevation (m)	445	105
Minimum elevation (m)	400	65
Watershed relief (km)	0.045	0.04
Relief ratio	0.006	0.006
Relative relief	0.002	0.002
Elongation ratio	0.671	0.823
Mean slope (%)	2.77	1.2
Drainage factor	0.7	0.60
Shape factor	2.83	1.878
Form factor	0.353	0.533
Circulatory ratio	0.64	0.765
Drainage density (km^{-1})	1.53	1.974
Ruggedness Number	0.079	0.079
Hypsometric Integral (H_{si})	0.58	0.5

The recent GIS interface programming languages are also flexible enough to accommodate advance hydrological models for real time data analysis and prediction. Therefore, the use of GIS in accounting the surface runoff, pollutant load, sediment losses, ground water information and preparation of a complete watershed water budgeting and optimal land use planning has been attempted by different researchers around the globe. The GIS environment assist in linking and interpretation of the attribute information of different watershed features representing the land use, soil type, ground water information etc., along with hydrologic, demographic and socio-economic data as shown in Fig. 6. Also, the interface programming language of advanced GIS software permits development of mathematical relations to generate the watershed responses and estimate the cost for construction of soil and water conservation measures (Razavi, 2002).

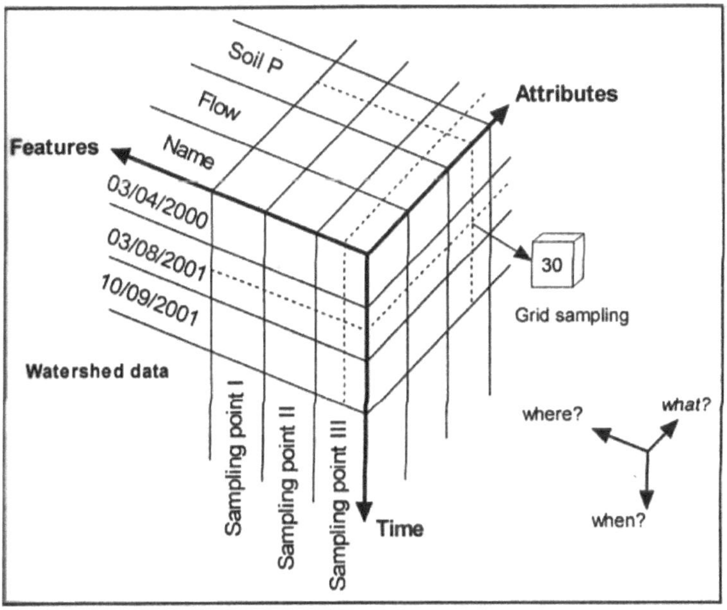

Figure 6. The GIS based watershed analysis for the temporal and spatial watershed feature classes and attribute data.

Remote Sensing (RS) in integrated watershed management

The image of the drainage basins or catchments areas acquired by the remote sensing satellite provide the preliminary step for acquiring the land topological, soil type, land cover/ land use information for subsequent management of the natural resources. In the area of satellite based remote sensing in the past, the first generation satellite IRS (Indian Remote Sensing)-1A and 1B were designed, developed and launched successfully during 1988 and 1991 with multi-spectral cameras with spatial resolution of 72.5 m and 36 m. respectively. Subsequently, the second generation remote sensing satellites IRS-1C and 1D with improved spatial resolutions of 70 m in multi-spectral and 5.8 m. in Panchromatic bands and a wide field sensor with 188 m resolution and 800 km. swathe, have been developed and successfully launched in 1995 and 1997, respectively. These satellites have become the principal components in the National Natural Resource Management System (NNRMS) and the data was used in various applications of integrated watershed management through acquisition of information on agriculture and soil, land form and land use studies, water resource, forestry, drought and flood monitoring and cartography. Especially IRS-1C/D data has been used for cartographic application up to 1:10,000 scale. These satellites also provide stereo pairs of imageries to get height information to an accuracy of approximately 10 meters. Recently, launching of CARTOSAT-1 (CARTOgraphic SATellite-1), a sun synchronous satellite, with two improved panchromatic cameras with 30 km swathe and 2.5 m spatial resolution, will generate more accurate digital elevation models (DEM). The DEM will be generated using the stereo photogrammetry techniques on the stereoscopic images acquired by CARTOSAT-1. Besides, these IRSs, there are also other satellite series such as LANDSAT, ASTER (Advanced Spaceborne Thermal Emission and Reflection Radiometer), SPOT (*Satellite Pour l'Observation de la Terre*), EROS, Quick Bird, ENVISAT, ERS, IKONOS and other Russian Satellites. Amongst these in panchromatic range the Quick Bird satellite provides the minimum resolution of 0.61m (Fig.7), where as IKONOS data records 4 channels of multi-spectral data at 4 m resolution and one panchromatic channel with 1 m resolution. The rest of the satellites records data at higher resolutions in both panchromatic and multi-spectral ranges. Moreover, uninterrupted collection of data can be achieved using Synthetic Aperture Radar (SAR) imagery of RADARSAT, which can circumvent cloud cover, rain, dust, haze and record data day and night.

Figure 7. The 61cm resolution image acquired through the Quick Bird Satellite over Tampa, Florida, USA.

Further, the RS images acquired from the satellites are available in form of digital data stored in optical storage devices, as orthophotos and as False Color Composites (FCC). All these raw data available from the satellite can be analyzed using the image analysis softwares (*viz.* ERDAS Imagine® (*www.erdas.com*), PCI Geomatica® (*www.pcigeomatics.com*), ENVI® (*www.rsinc.com*), ER Mapper® (*www.ermapper.com*), Image Analyst® (*www.intergraph.com/ianalyst*), TNTmips® (*www.microimages.com*) , ILWIS® (*www.itc.nl/ilwis*) etc.). However, TNTmips® and ILWIS® are also used as both the RS image analysis and GIS data analysis tools. However, an important feature of the RS image analysis is to acquire the ground truth information corresponding to the digital signatures in the RS imageries. The acquisition of the ground truth information is essential for generating more accurate feature classes for generation of geo-spatial information. The generated DEM of the land surface can be used to delineate watershed and the geo-processing operations of the land use, soil type and surface and ground water availability will yield scenarios of efficient soil and water resources management.

Image data have been used as a primary source of natural resources information in thematic mapping which in turn is utilized for various hydrological studies. The remote sensing data provides synoptic view of a fairly large area in the narrow and discrete bands of the electromagnetic spectrum at regular intervals. The space borne multi-spectral data enable generating timely, reliable and cost effective information on various natural resources, namely surface water, ground water, land use/ cover, soil, forest cover and environmental hazards (such as: water logging, salinity and alkalinity, soil erosion by water etc). For many hydrological purposes, remote sensing data alone are not sufficient and need to be merged with data from other sources. Hence a multitude of spatially related and geo processed data concerning topography, rainfall, evaporation, vegetation, geomorphology, and soils have to be considered including the social and economic data related to watershed developmental programmes, irrigation, locations and types of tube wells, rain gauges and silt monitoring stations over watersheds. The combination of RS image analyst and GIS data provides an extremely useful technology for considering the interaction between spatially distributed resources. Watershed based analysis to predict surface runoff and sediment losses from a given amount of precipitation have been performed by the researchers using the capabilities of RS image interpretation, hydrological modeling tools and geo processing operations of GIS. The most commonly used methods attempted by the researchers in generation of surface runoff is by curve number (CN) approach and prediction of sediment yield is attempted by using universal soil loss equation in its modified and revised forms. Also, the watershed health indices are developed using the RS image bases analysis and GIS.

Decision Support Systems in IWM

The field of decision support systems (DSS) is considered as a third generation of computer-based applications, after transaction processing (TPS) or 'efficiency' systems, and management information systems (MIS) or management reporting systems. Decision support systems evolved from the concept proposed by Gorry and Morton (1971), who explored the concept of structure in decision making by constructing a matrix showing the interaction between the level of management and the decision making structure at each level.

DSS is also defined as a computer-based systems that will aid or support decision makers to solve unstructured problems through direct interaction with data and analytical models. It is a mechanism that facilitates complex decision-making. DSSs combine the methodology of decision analysis (derived from systems analysis) and the information component of decision support into a process that links directly to the decision-making process. In general, DSS is defined as "contemporary jargon for an integrated approach to the age-old problem of helping people make better decisions". The essence of developing a DSS is therefore the integration of information and methodologies from different sources for the purpose of assisting in managing complex systems. It is an integrated assembly of models, data, interpretive routines and other relevant information that efficiently processes input data, runs the models, and displays the results in an easy to interpret format. The common goals of DSS methodologies in this context are better information, greater objectivity and higher efficiency (Sarangi et al., 2004a). The components of a DSS for IWM is shown in Fig. 8.

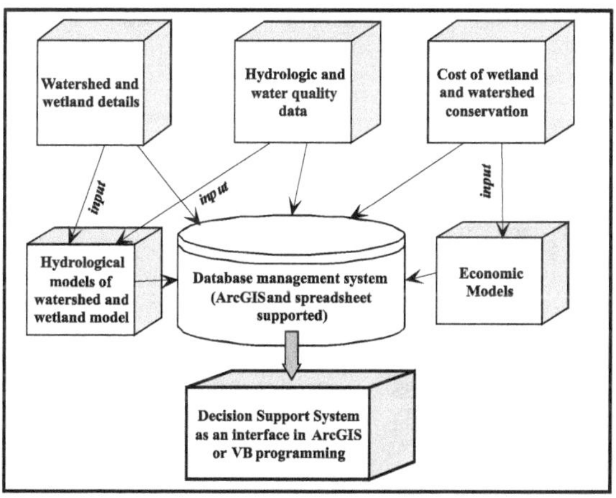

Figure 8. The components of a DSS for Integrated Watershed Management (IWM)

During last decade, attempts and approaches to develop decision support systems and expert systems in the area of IWM are numerous. The DSS can be developed as stand alone software incorporating the conceptual system flow logic using the programming languages with dynamic linkage to the data base management system. DSS can also be developed as an interface with GIS and RS softwares using the built-in macro programming languages linking the GIS feature classes, hydrologic models and interpretive routines. One of the first attempts to create a prototype multi-objective decision support system (MODSS) was undertaken by Yakowitz *et al.* (1993). They introduced MODSS which combined the modeling of hydrology, erosion, nutrient and pesticide transport as well as crop growth and economics with the multiple objectives of reducing surface and groundwater pollution while maintaining or improving farm income. All components where analyzed using different scenarios. One of the main focus was upon the involvement of the practitioner in the decision making process. MODSS allowed the user to set up alternatives, choose the decision criteria, and accept the customized decision mode settings. The decision results were presented graphically as well as in tabular form. Yakowitz *et al.*(1993) claimed that if differences between the scenarios existed, they would be obvious in the decision support system output, therefore a support routine to compare the results of the simulations analytically was lacking. All these DSS's were developed as stand alone software using the visual C++ and visual basic (VB) programming languages.

Srinivasan and Engel (1994) described the development of a decision support system for assessing agricultural non-point source pollution at the basin scale. It is a system for improved model application such as model input and output coupling with Geographical Information Systems. With respect to the decision making process an interface for the decision-makers was still lacking. Furthermore, the used models (SWAT, SWRRB) divide the investigation area into sub basins by the homogeneity of soil, crop, and topographic feature. This stands in contrast to watershed scale models like AGNPS or

ANSWERS that use a raster based approach to determine the input parameter, which will allow a better area differentiation, since most geographical areas are usually quite diverse in their soils, crop or topography. Chen et al (1996) explained a software package being developed for decision-making, although they did not call it as a DSS. However, it assists the decision-maker by providing an evaluation system to convert quantitative model output into decision variables. The system was applied to the Houston River only. Jamieson and Fedra (1996a; 1996b) reported the use of a European decision support system called WaterWare for integrated watershed management including water supply, land drainage, hydropower generation, effluent disposal, recreation and other amenities. In addition to earlier DSS approaches, the authors emphasized the importance that a DSS should be open to the public. Consequently, their design of the DSS provided an information module that could be understood by the non-experts, both concerning basic information, data and model output interpretation. Andreu *et al.*, (1996) presented the AQUATOOL DSS for integrated basin management. Again AQUATOOL was designed to mainly support the modeling part. The decision support itself was not a part of the system.

Another example of DSS is WATERSHEDSS described by Osmond *et al.*, (1996), which had most of the desired components incorporated. WATERSHEDSS (WATER, Soil, and Hydro- Environmental Decision Support System) was designed to help watershed managers and land treatment personnel identify their water quality problems and select appropriate best management practices. It still lacks the ability to compare the best management practices against one another and evaluate the results. Lohr and Ostrowski (1996) developed a DSS for improved ecological operation of reservoir systems, this research lead in 1997 to the design of the STEEL - GDSS (Science, Technology, Sustainable, Ecology for Urban Economy Storm water Law Treatment Group Decision Support System And Information Network) in cooperation with the University of Guelph. Again this system was aimed at large scale operations. Recently, Sarangi et al., (2004a) developed a DSS for generating alternative scenarios on use of soil and water conservation measures for integrated watershed management based on the slope, annual sediment load, land capability classes and hydrologic soil groups of selected location over the watersheds. The DSS is capable of providing sediment control solutions not only for small watersheds but also for larger drainage basins, by dividing the basin into smaller watersheds. The DSS was validated for a watershed on the Caribbean island of St. Lucia and used to suggest measures for a 10° slope under specific soil type, sediment loss and LCC conditions. The measures proposed included bench terraces, graded contour bunds, conservation ditches, concrete chute spillways, diversion dams and conservation cropping systems. The measures actually adopted on-site were conservation ditches, graded contour bunds and conservation cropping systems, a close parallel to the DSS's proposed measures. On slopes ranging from 5-55°, imple-mentation of the suggested control measures resulted in a 34-37% reduction in soil loss on the watershed. The operational window with the input and out put decision boxes are shown in Fig 9. Therefore, it is evident that the GIS, RS and GPS tools can be efficiently used to process all the land water and vegetation related features and other demographic and socio economic information within a single environment resulting in development of a meaningful DSS. The system architecture achievable through GIS interface and stand alone programming software linking different data based and GIS generated thematic maps are presented in Fig. 10.

Use of Geospatial tools for development of a Ground Water Information System (GWIS)

A support system to estimate the ground water budgeting within a watershed system and locate possible recharge zones to augment ground water was developed within ArcGIS® by linking different geo-spatial information. The flow chart of the Ground Water Information System (GWIS) interface is shown in Fig. 11. The basic functionality of GWIS interface was to estimate the ground water availability, based on the standard techniques viz. water level fluctuation method and infiltration

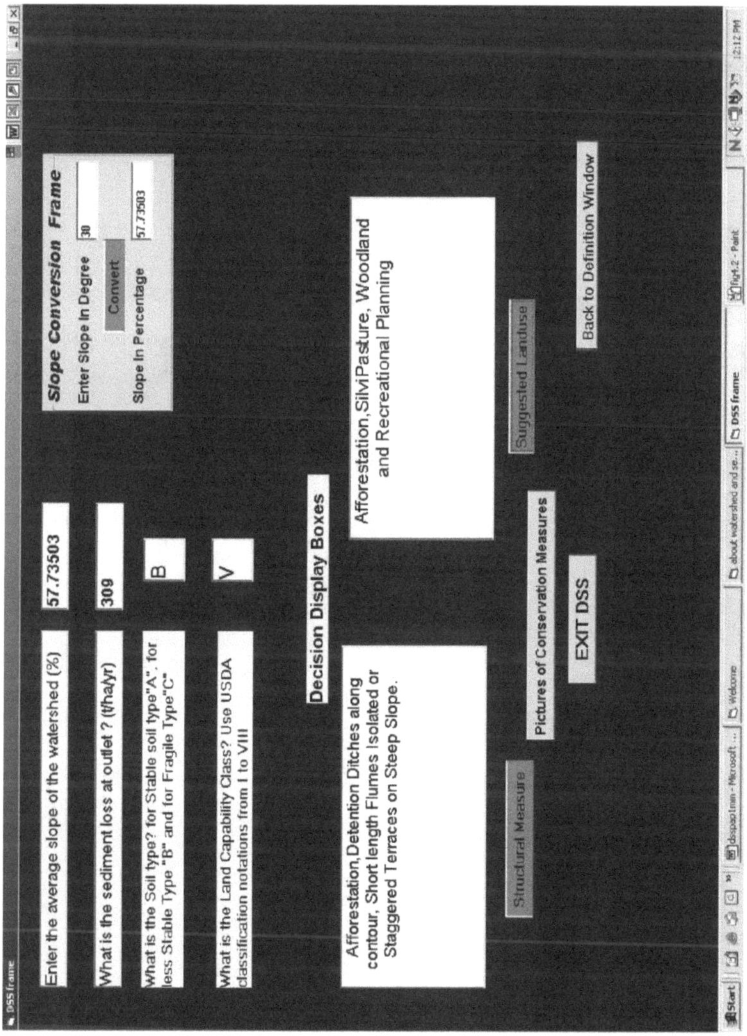

Figure 9. The main DSS frame displaying the suggested measures of St. Lucia Watershed

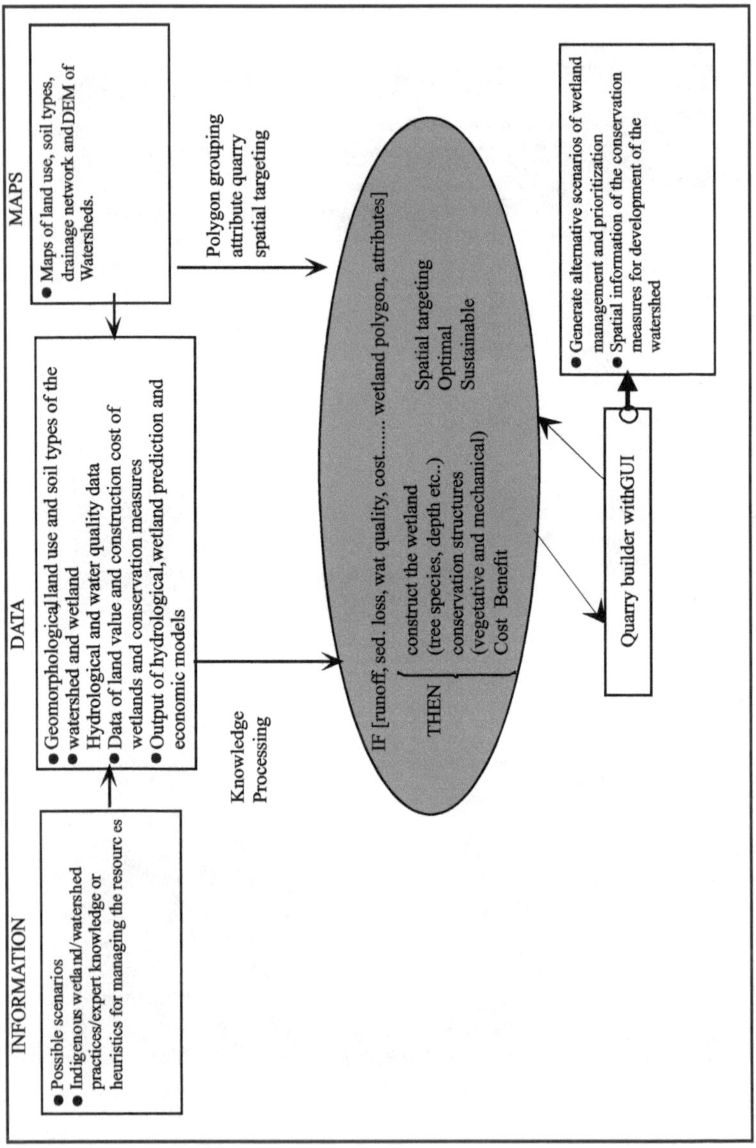

Figure 10. A typical architecture of the DSS for IWM

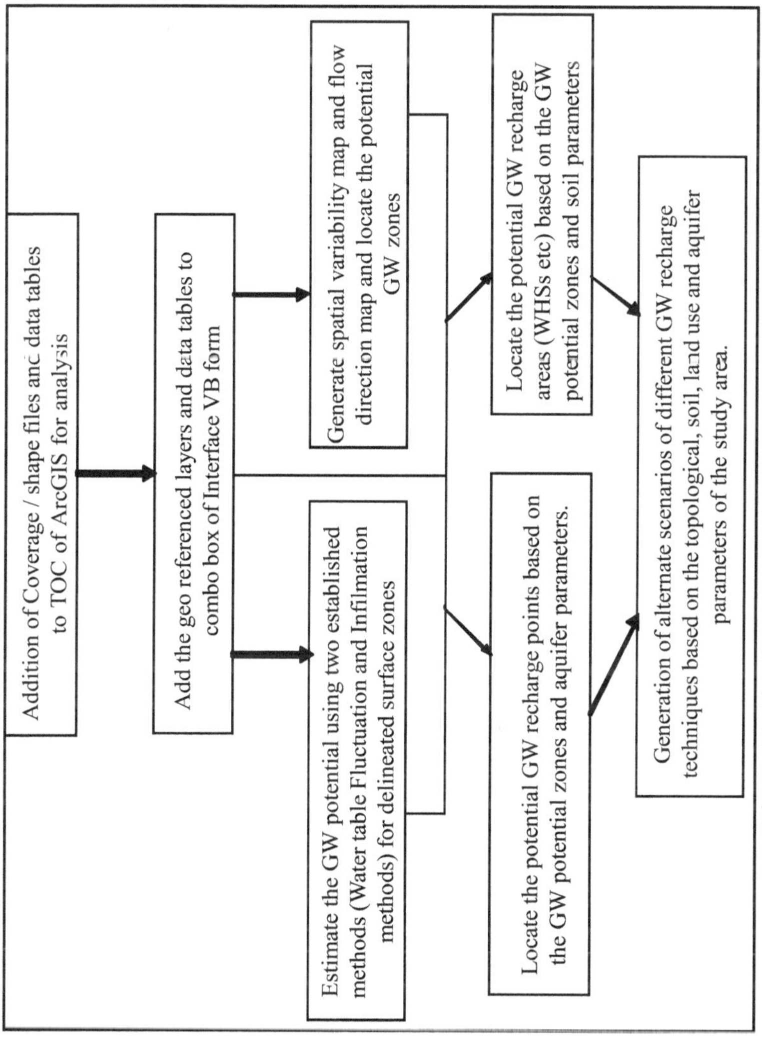

Figure 11. The flow chart of GWIS interface with ArcGIS®

method, estimate the spatial variability and flow direction of ground water and location of the potential recharge zones. The interface developed using VBA, the built-in macro of ArcGIS®, which can automate repetitive tasks or create complete applications with relative ease within the GIS environment. The programming syntaxes and logical expressions of GWIS performs map manipulation, data analysis and display within the GIS environment. The visual basic forms with button controls are developed to link the interface with each distinct activity starting from activation of GWIS interface (Fig. 12). While developing the interface, the flow logic as detailed in Fig. 11 was used for linking different data sources acquired by using GPS, digital spreadsheet data and processed using GIS and RS tools. The Interface requires the point, line and polygon feature classes and attribute tables of the study region related to the surface topology (contours), land use, soil physical properties, well log information, water table depth fluctuations, rainfall runoff information, hydrologic soil groups and aquifer parameters (specific yield, storage coefficient, hydraulic conductivity, Transmissibility etc.). The command buttons of the GWIS interface (Fig. 12) imports the contour, isobath, land use and soil type thematic maps and the spreadsheet data related to aquifer details and the soil physical parameters to the Table of Content (TOC) of ArcGIS®. Other operating command buttons perform geo-processing operations to locate ground water flow accumulation, estimates the ground water potential of the watershed system, extracts aquifer details, soil properties and relates with the topological features to locate the potential recharge points and area for construction of recharge shafts and water harvesting structures. The GWIS also generates different ground water recharge techniques based on the watershed morphology and the geo-processed information of water movement within the surface. The developed interface is user friendly and flexible enough to incorporate the modification and fix the operational bugs and errors. The generated output of ground water recharge locations and annual ground water balance can be displayed on the arc map and the attribute information are written to specified output file locations.

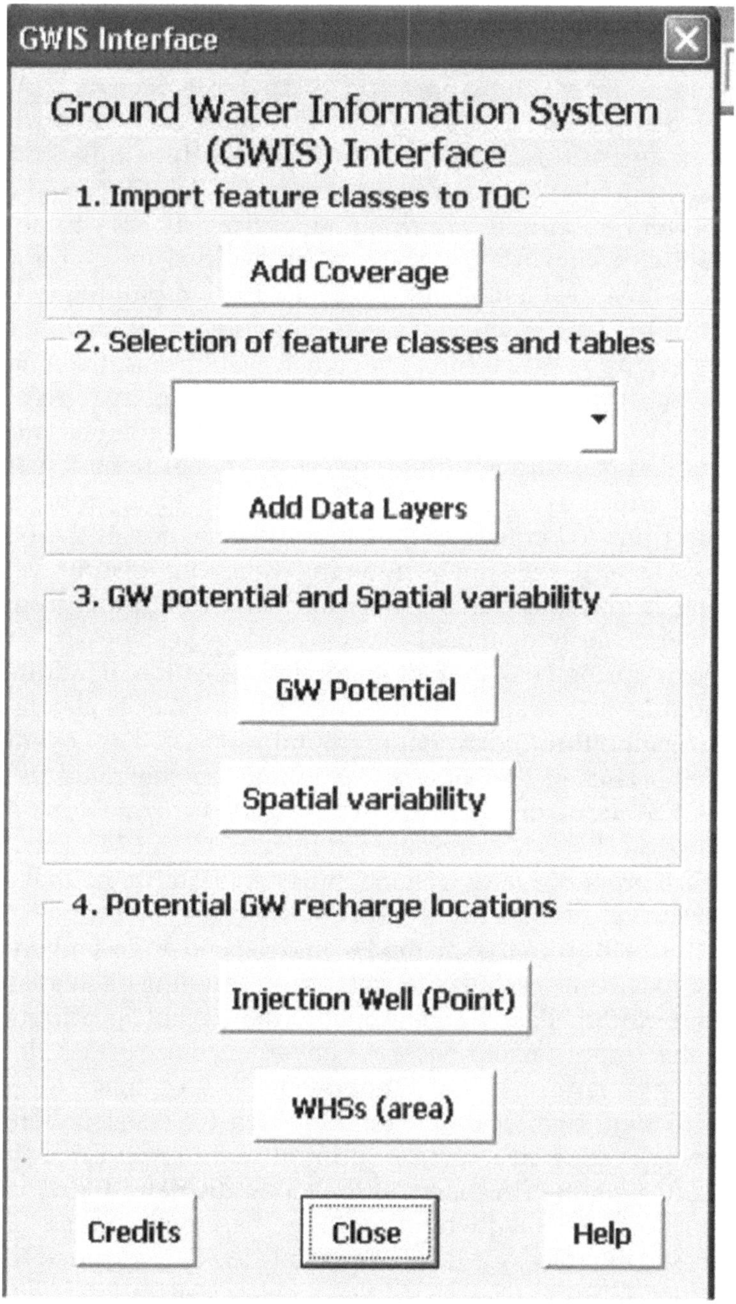

Figure 12. The GWIS interface with command buttons

The developed GWIS interface can be applied to any watershed locations with the specified input data requirement. In this chapter, the operation of the interface for IARI watershed (473 ha) is discussed. The delineated watershed of Indian Agricultural Research Institute (IARI) campus, New Delhi, is situated between the latitudes (28^0 37′ 22″ N and 38^0 39′ N) and longitudes (77^0 8′ 45″ E and 77^0 10′ 24″ E) at an average elevation of 230 m above mean sea level and comprises of farm area, residential complexes and office buildings. The relief of the watershed is only 23m and about 290 ha area is under extensive agriculture. The climate is subtropical, semiarid with hot dry summer and cold winter and falls in Agro-eco-region-IV (semiarid with alluvium derived soil). The mean annual temperature is 24^0 C. The mean annual rainfall is 710 mm, of which as much as 75 % is received during monsoon season (June to September). Most of the irrigation at the IARI farms is being carried out by pumping water through tube wells, storing it into tanks and delivering it through underground pipelines to individual fields. Water shortage becomes a crucial problem during peak demand period of irrigation. In addition to the tube well water, water from *Yamuna* River is also used for irrigation through an underground pipeline. The coverage feature classes of the soil type and land use generated using the GPS acquired latitude -longitude information and topological sheets are presented in Figures 13 and 14. The soil physical properties and well log information indicated that the watershed is suitable for groundwater recharge. However, the selection of the recharge method would depend on cost involved and capacity to recharge a maximum amount of available surface runoff. Based on the methodologies of locating the potential zones related such as convergence of surface flows (topology), land use, soil type, availability of space for the construction, two locations are identified for construction of water harvesting structures (Fig. 15). It was also revealed that, the construction of recharge shafts along the natural drainage channel of watershed would recharge the ground water. The discussed interface was developed using the GPS acquired land positions, GIS and RS assisted feature class generation and use of VBA protocol of ArcGIS® is coding the support system. This interface needs to be applied to other watershed regions to ascertain its performance and wide acceptability. In another example of development of support systems using all these geospatial tools, Sarangi *et al.,* (2005) developed a DSS for watershed hydrological responses using the ArcGIS® and VBA programming language to generate alternative scenarios of runoff and sediment losses from watersheds based on their geo-morphological characteristics.

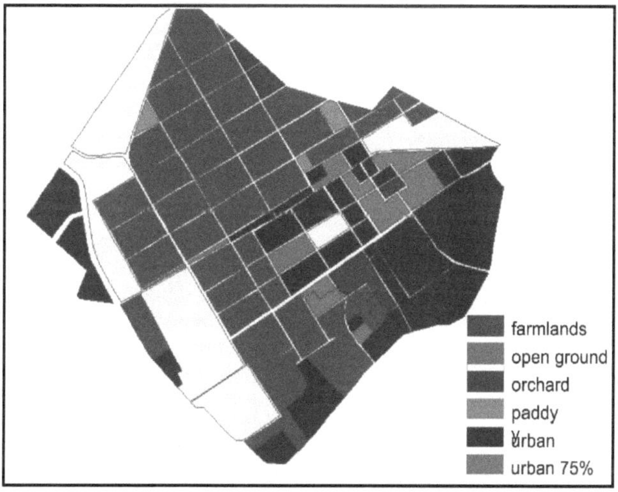

Figure 13. The land use map of IARI prepared using ArcGIS® and GPS

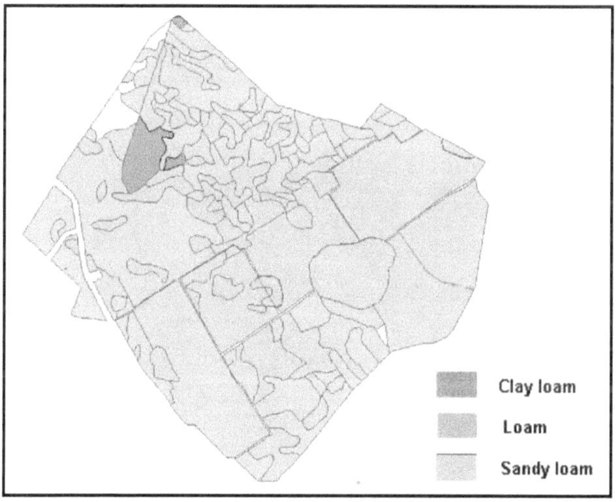

Figure 14. The soil map of IARI prepared using ArcGIS® and GPS

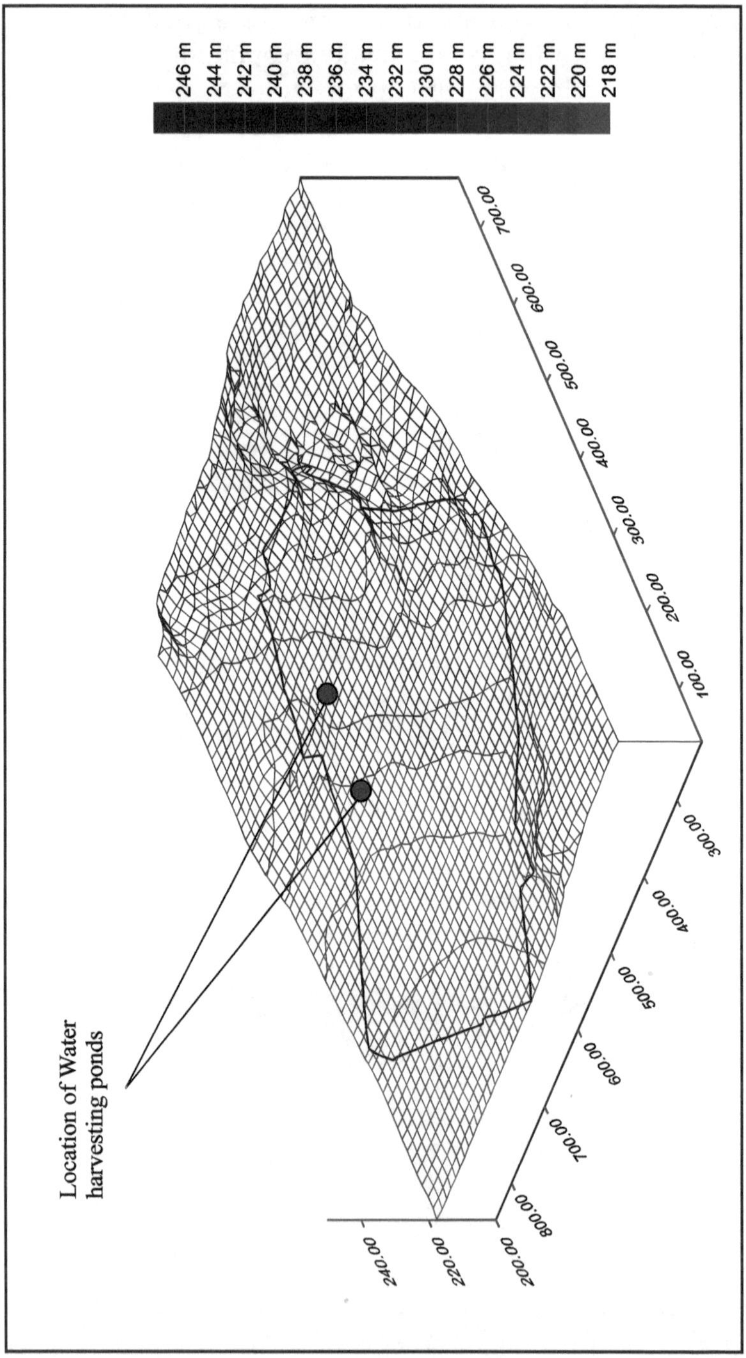

Figure 15. GWIS generated location of potential water harvesting zones in IARI watershed

Conclusions

The belief that watersheds make a sound basis for water resource planning and management is not new, as evidenced by waves of scientific, policy, and public interest going back as far as the 1930's. Yet after many years of high expectations, the nation is still struggling to find ways to implement integrated management at the watershed level. Much of the science and technology needed to provide the underpinnings necessary for integrated water management already exists. Numerous scholarly reports have highlighted the potential benefits to be gained from a watershed approach. But we have fallen short in turning our understanding of watersheds and the benefits of integrated management into action. Therefore, faster data processing and proper understanding and analysis of the hydrologic process, soil and vegetation resources in watersheds can be accomplished through the use of advanced geospatial techniques, which in turn can assist in generation of scientific methods leading to efficient IWM.

The geospatial tools such as GPS, GIS and RS can be effectively applied to data generation, analysis and interpretation leading to development of a exhaustive digital information of the land features, soil and water resources. The detailed information generated by using the geospatial tools can be used as an input to the hydrologic and water quality prediction models for quantification of runoff, sediment yield and pollutant loads from the watersheds. The operation of the models under different input conditions will generate alternative output scenarios and all these information can be synthesized to develop a DSS. The developed support systems collates all the process information and assists in estimation of the water yield potential of a watershed and also suggests the best management practices to conserve water, arrest soil loss and prevent contamination of the ground water and downstream rivers. The suggested measures can be implemented within the watershed systems for attaining sustainability in agricultural production.

References

Andreu, J., Caprilla, J., Sanchis, E. (1996) AQUATOOL, a generalized decision-support system for water resources planning and management. *Journal of Hydrology*,177: 269-291.

Borah, D.K. and Bera, M. (2004). Watershed-scale hydrologic and non point-source pollution models: review of applications, *Trans of ASAE*, 47(3): 789-803.

Chen, C.W., Herr, J., Goldstein, R.A., Sagona, F.J., Rylant, K.E., Hauser, G. E.: (1996) Watershed risk analysis model for TVA's Holston river basin. Water, *Air and Soil Pollution*, 90: 65-70.

Diplas, P. (2002). Integrated decision making for watershed management: Introduction, *Journal of American Water Resources Association*, 38(4):893-894.

Garg, V., Chaubey, B.E. and Haggard, B.E. (2003). Impact of calibration watershed on runoff model accuracy, *Trans. of ASAE*, 46(5): 1347-1353.

Gorry, M. A. and Scott, M. S. (1971) A framework for management information systems, *Sloan Management Review*, vol.13, Massachusetts Institute of Technology: 55-70.

Haan, C.T. (1995). Models and Decision Making in Uncertain Environments. Proceedings from Workshop on Computer Applications in Water Management, Fort Collins, CO. Water Resources Research Institute, Fort Collins, CO.

Jamieson, D.G., Fedra, K. (1996a): The WaterWare Decision Support System for river basin planning: 1. Conceptual design. *Journal of Hydrology*, 177: 163-175.

Jamieson, D.G., Fedra, K. (1996b): The WaterWare Decision Support System for river basin planning: 3. Example applications. *Journal of Hydrology*, 177: 199-211.

Kerr, J and Chung, K. (2001) Evaluating watershed management projects. *Water Policy* , 3: 537-554.

Lea, N.L. (1992). An Aspect Driven Kinematic Routing Algorithm. Overland flow: Hydraulics and Erosion Mechanics, A.J. Parson and A.D. Abrahams, (eds.), Chapman and Hall, New York.

Lohr, H. and Ostrowski, M.W.: (1996): Planungswerkzeuge für den ökologisch orientierten Talsperrenbetrieb, Wasser Abwasser Praxis, Mai: 22-25.

Martz, L.W. and Garbrechet, J. (1992) Numerical Definition of Drainage Network and Subcatchment Areas from Digital Elevation Models, *Computers and Geosciences*, 18(6):747-761.

Osmond, D.L. *et al.*, (1996) Watershedss: A decision support system for watershed-scale non-point source water quality problems, in Proceedings of the First International Conference on Multiple Objective Decision Support Systems for Land, Water and Environmental Management: Concepts, Approaches, and Applications, Lewis Publishers, New York: 241 –250.

Sarangi, A. Madramooto C.A. and Cox, C. (2004a) A Decision Support System (DSS) for Soil and Water Conservation Measures on Agricultural Watersheds, Land Degradation and Development, 15(1): 49-63.

Sarangi,A, C.A. Madramootoo, D.K. Singh and A.K. Singh (2004b) Performance of GIS interfaces in Watershed Delineation and Stream Network Generation from DEM, *Journal of Indian Society of Agricultural Engineering*,41(2): 41-48.

Sarangi, A., D.K. Singh, A.K. Singh and C.A. Madramootoo (2005) Development of DSSWHR- An ArcGIS Based Decision Support System for Watershed Hydrological Responses. In: Proceddings of International conference on Hydrological Perspectives of Sustainable Development, 23rd to 25th Feb 2005, IIT Roorkee, Uttaranchal, India, Vol. I: 20-31.

Srinivasan, R. and Engel B.: (1994) A spatial decision support system for assessing agricultural non-point source pollution, *Water Resources Bulletin*, 30(3): 441 - 452.

Wilson, D.J. and Droste, R.L. (2000). Design considerations for watershed management decision support systems, *Water Quality Research Journal of Canada*, 35(2): 163-188.

Yakowitz D. S., Stone J. J., Lane L. J., Masterson P. J., Abolt J. and Imam, B. (1993) A Decision Support System for Evaluating the Effects of Alternative Farm Management on Water Quality and Economics, *Water Science and Technology*, 28(3): 47-54.

❑❑❑

GEOINFORMATICS APPLICATIONS IN AGRICULTURE, 2007
Eds. : A.K. Singh and U.K. Chopra; pp. 245-269
New India Publishing Agency, New Delhi (India)
E-mail : newindiapublishingagency@gmail.com
Web: www.bookfactoryindia.com

9

Geoinformatics for Enhancing Productivity in Flood Plain Agro-ecosystem

R. N. Sahoo, R.K. Tomar, I.P. Abrol, R.K. Gupta, D. Chakraborty, V.K. Sehgal & N. Kalra

Key words : Flood Plains Agro-ecosystem, GIS, RS, Spatial Cropping Pattern, Radiometric Normalization, NDVI, Agricultural Systems.

Introduction

Of late, evidence is accumulating to suggest that high productivity and growth rates achieved during the Green Revolution era in the well-endowed regions of the country (*i.e* North West India) are no longer being sustained to meet the needs of still increasing population in the country. Past growth sources have nearly exhausted and there is also no scope for expansion of crop land. Indeed pressure for competing land uses will likely reduce area under food grain crops. Shrinking natural resource base (per capita availability of land and water devoted to agriculture), declining quality of resources and environmental degradation issues will imply increasing threat to our ability to meet the basic needs of the growing population of the region unless we shift our effort to potential regions like flood plains of the country for improving productivity.

The flood plain riverine lands of India covering approximately 42 m.ha though most endowed with natural resources and has all the potential of becoming rich, is still economically backward. The reason behind this paradox of extent of poverty in the midst of richness of resources is that the natural resources of the region are not being productively harnessed into the process of the economic development and thus are not delivering their potential and commensurate benefits to the people. The Green Revolution technologies evolved during of 60's and 70's had little or no impact in this region of the country. Undesirable spatio-temporal distribution of flood in these areas keeps agriculture under uncertainty and risks. These are also the areas where the poorest are concentrated and agriculture largely continues to be more of subsistence. The nature of agricultural problems are highly variable due to great diversity in the production system as also in the socio-economic conditions of the communities requiring location specific approaches to address the problems and enhance the productivity.

Recent development in geoinformatics particularly in the field of Satellite Remote Sensing, GIS and GPS technologies have special advantage in terms of generating state-of-the art informatics - capturing the variability, vulnerability and dynamism of flood plain agricultural system. These information would enable us to provide valuable scientific insights into the factors contributing to the low productivity which in turn would form the essential ingredients to evolve the effective strategies to enhance it.

Flood Scenario in India

India, on account of its geographical position, climate and geological setting, is the worst-affected theatre of disaster in the South Asian region after Bangladesh and accounts for one fifth of global death count due to flood. The most clinching evidence of floods having increased as a physical phenomenon comes from the increase in the flood affected area which increased from an annual average of 6.48 m.ha in the 50s to over 9 m.ha in the 80s. The Rastriya Barh Ayog (RBA, 1980) or National Commission of floods found that the country's flood prone area which was 25m.ha during the 60s went up to more than 42 m.ha in 80s. The most frequent flood prone areas are the Brahmaputra basin in Assam, the central and lower portion of Ganga basin in Uttar Pradesh, Bihar and West Bengal and the deltaic region of Orissa (Fig.1). These are spiraling out of control with increase in flood frequency, causing more and more injury, disability, disease and death, economic and social burden of an already impoverished nation.

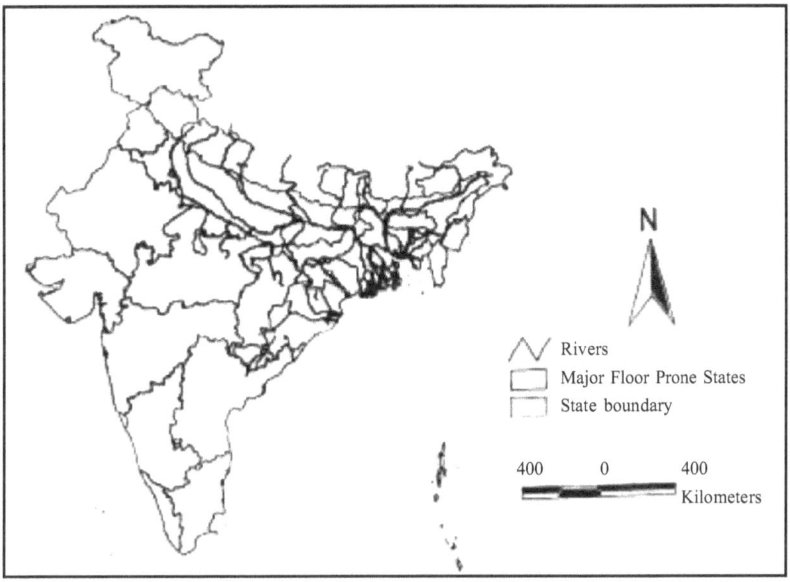

Figure 1. The most flood prone areas of India indicated with river basin.

Forces and their interplay for Flood Severity

Flood is a combined result of natural, ecological and anthropogenic factors which is shown through a conceptual framework in Fig. 2. Drought and floods are regular features due to erratic distribution of rainfall and other climatic parameters while one part of the region is under the grip of severe floods due to excessive rainfall; the other part suffers from drought due to poor rainfall. River Ganga and Brahmaputra and their tributaries emerge from mountains, which are not only highly erodible because of change in land use and land cover but its various geophysical phenomena such land slide, seismic activities, glacier melting etc (Hofer, 1993). The Himalayan rivers drain through its steep slopes where the

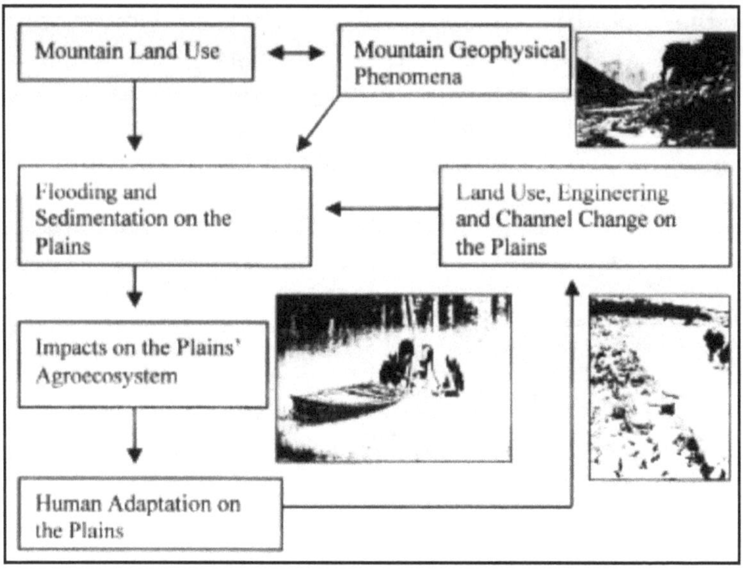

Figure 2. Forces interplaying for severity of flood and its impact on agro-ecosystem.

rainfall is copious and during monsoon they carry discharges much in excess of their channel capacity along with huge quantities of sediments from its erodible slope. On emerging from the hills, their slope remains fairly steep which become gradually flatter down in the plains and is associated with decrease in velocities and consequent deposition of more and more silt and sediment. The heavy sediment deposition on riverbeds particularly in the flatter reaches reduces the carrying capacity for flood discharge and the water overtops the bank inundating the adjoining areas. Sometimes the river erodes their banks and change courses abruptly, opening up new spill channels and thus cause considerable damage to the adjoining areas. Floods are also caused by excessive rains in river catchments. The magnitude of flood depends upon the nature and extent of rainfall in the catchments. Mostly over spilling of the river on its banks causes flooding. This is generally due to excessive rainfall, combined with inadequate channel capacity. Damage due to flood in the plains has increased exponentially not only because of increase in flood severity but majorly due to increase in human settlements in these regions over time. To avoid such damage, some of the human adaptations in terms of changing land use pattern particularly for agriculture and built up, engineering structures to protect from flood has taken place. However, these in turn act as one of the factors causing flood. Raising of embankments reduces river flow allowing sediment deposition right near the embankment. For example, the master drain in Ganga remains in high stages during monsoon and blocks the tributary drainages and results in drainage congestion in upstream reaches. Over spilling in the river also occurs due to obstruction in the riverbed, aggradations of the riverbed on account of heavy silt load brought down during monsoon. As the agricultural lands are protected from flood by embankment, no more silt deposition is there and thereby reducing the fertility of the land. Farmers of the protected land no more have bumper harvest of the crop compared to that taken after flood. It is still a debatable issue to be evaluated whether making flood protection structures is a boon or curse.

Flood Plain Agro-ecosystem

Flood plains are formed by fluvial soils and its development is directly related to (i) the existing fluvial processes, (ii) the mode of sediment transportation and (iii) finally the deposition of transported materials under a set of variables, which influence the mode of deposition of sediments in the flood plain and contribute appreciably to the genesis of fluvial soils are

- Geomorphology of the flood plain
- Operative fluvial processes
- Intensity and periodicity of flood
- Depth, flow velocity and direction of flood
- Distance from main stream and channels controlling the fluvial processes
- Nature of river beds (straight, meandering or braided)
- Vegetation

Various groups of workers have attempted to classify the flood plains. Some common definitions for the classes are

- Diaralands - These are situated in between the natural levees that get inundated for different periods of time and are periodically eroded and formed by the meandering, braiding and the changing course of the rivers (Fig. 3).
- Tal lands - Beyond the natural levees, there are bowl shaped depression known as backwaters which though inundated are not subjected to erosion. These are also called as wetlands.
- Chaur- Chronically flood affected lands. Mostly unstable on the riverside, but away from the river, these become stable.

Figure 3. Schematic diagram showing diaraland in a river system (*Source:* CSE, 1991).

A Technical Group (Anonymous, 1974) on diara areas estimated the diara land for various river systems on the basis of 3 by 5 kms of diara for every kilometer length of river which has been estimated to be approximately 2.90 m.ha spread mainly in the states of Uttar Pradesh (1.5 m.ha), Bihar (0.9 m.ha) and

Assam (0.24 m.ha). How ever, the estimates of diara areas in West Bengal, Orissa and other states having major river systems are not available. Diaralands are characterized by absence of distinct horizons and undulating topography. There are three distinct physiographic situations encountered in diara lands. The highland diara is subject to inundation only during high floods, which may occur once in five years. While medium land diara may be flooded once in three years and low lands are subject to flooding every year or may be flooded more than once in a year resulting 50 to 80 % to crop damage. Flooding period does not exceed more than a month in late August-September in a year in upland diara, extends from one to three months (July to Oct) in medium diara and prolongs more than three months in low land diara. Low land diara exhibit poor drainage due to lack of outlet and high water table conditions. Such areas have the flood problem during monsoon (June to September) which substantially reduces the growing season while during winter and summer, crops are raised on residual moisture and very often moisture stress reduces crop yields because there is no assured irrigation (hardly 3 to 10 % area is irrigated).

Flood Plain agro-ecosystem differs widely from that of the other plains because of its distinct geophysical composition, the recurrence of annual floods in the rivers and its erosion. Cropping pattern is based on subsistence farming. Farmers follow two main cropping seasons in the eastern part of India viz. *kharif* (June-July to September-October) and *rabi* (October-November to March-April). Often the crops planted during June are lost due to heavy floods and it may be possible to harvest only those crops that are planted during October-November (Anonymous, 1981-1983). Thus, rabi is the only season where diara farmers are assured of a good crop. But 90 % of such areas do not have irrigation facilities and hence rainfed farming is the only alternative. Mixed cropping is the traditional system as an insurance against complete crop failure. In U.P, sugarcane, rice, pigeon pea, sorghum and jute are raised during *kharif* followed by crops like wheat, lentil and mustard during *rabi*. In Bihar, land remains fallow during *kharif* while crops like maize, sweet potato, wheat, mustard and pointed gourd are raised during *rabi*. In Assam, it is mostly rice (Ahu, Bao and Sali), which is taken as a sole crop or mixture of varieties of different durations in an agricultural year.

Constraints and Challenges

The problem of improving agriculture in flood plains are rendered complex due to uncertainty and risks associated with undesirable spatio-temporal distribution of flood, little or no possibilities of building permanent assets and focus on basic livelihood needs, uncertainties of land ownership and tenurial system, lack of communications and unique interaction of biophysical and socio-economic factors adversely affecting the ability of farmers to adopt improved practices that will contribute to stability or enhanced incomes. Demarcation of owners' or tenants' field after flood recession is a major cause of social conflict and these areas are hotbed for anti-social activities. Flood plain inhabitants are forced to migrate seasonally or permanently. Migratory poor population cannot have various types of livestock though it acts as cushion like a liquid asset. People develop an attitude of survival strategy rather than an outlook of profit maximization. Food security gets high priority and profit-making agriculture gets back seat.

Developing alternate strategies and appropriate technologies for agricultural development demands a much better understanding of the embedded unique interaction of biophysical and socio-economic characteristics that influence the lives of the inhabitants of flood plains. The technological needs are characterization of flood plains, forecast of flood hazard and timely approaches to minimize its ill effects, minimizing risk of floods, understanding the Farming System situation in flood plain for developing and promoting intervention that will stabilize and enhance income of farmers and understand migration and survival pattern for developing coping strategies that integrate conventional wisdom and scientific advances for sustainable growth and development of flood plain agriculture and socio-economic condition of the poor.

Role of Geoinformatics

Strength of remote sensing and GIS lies in reaching out and unfolding the various linkages and the underlying factors existing between state of the natural resources and environment and incidence of disaster induced low productivity and food insecurity. The advantage of remote sensing technology has been its inherent capability that provides synoptic and repetitive high resolution imageries over large areas and thus maps out the spatio-temporal variability of floods and crops. Geographic Information System (GIS) establishes dynamic linkages between biophysical factors with socio-economic variables for analyzing low productivity through modeling and simulation. A conceptual frame work for use of remote sensing and GIS for enhancing livelihood of the flood plain agro-ecosytem is given in Fig. 4 which was executed in Eastern Gangetic flood plain of Bihar (Sahoo *et al.,* 2002).

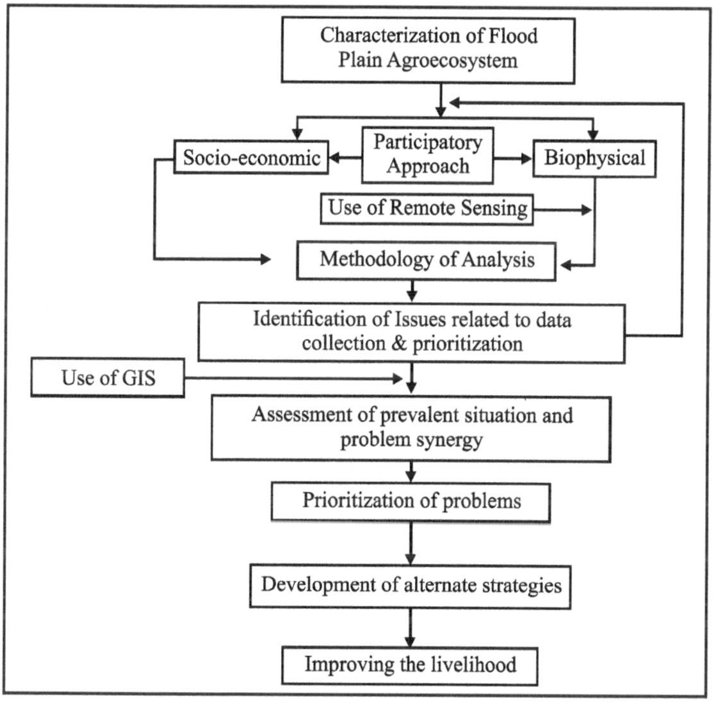

Figure 4. Conceptual framework for improving livelihood of flood plain agro-ecosystem.

Targeting the vulnerability

Vulnerability of agriculture to flood is well known. Remote sensing captures the vulnerability through near real time flood monitoring, its impact assessment and forecast modeling. Mapping of flood-inundated areas is one of the most successful applications of satellite remote sensing and this is routinely operational in the country since 1988. Because of the unique spectral signature, it is possible to map areas under standing water, areas from where flood water receded, submerged standing crop areas, sand casting of agricultural lands, breaches in the embankments, marooned villages and towns etc. Using multi-date satellite imageries, the extent of damage due to crop loss, destruction of infrastructure facilities etc can be assessed. Satellite remote sensing and GIS techniques have been integrated (Venkantachary *et al,* 2001) in Brahmaputra river basin to provide information on flooded area and damage to cropping lands, roads and rail tracks. Global Positioning System (GPS) is being used to aid in the development of Digital Elevation Model (DEM) for a flood prone area in Andhra Pradesh, to enable assessment of spatial inundation at different water levels in the river. Satellite derived land use and land cover and ancillary ground based socio-economic data is assessed to provide location specific flood warnings. Remote sensing data are evaluated for integration with existing forecasting models. Also microwave data from RADARSAT is used in conjunction with optical data to overcome the limitation of cloud cover. A proto-type system consisted of comprehensive database design, space-based near real time monitoring tools, modeling framework, networking and user interface was synthesized by Indian Space Research Organization (ISRO), Bangalore in 1998. The date wise flood inundation and recession pattern during the event year (2003) and its spatial extent during 1988 to 2003 in Eastern Gangetic Flood Plain of Bihar is given in Fig. 5. The year 2003 was flooding event year for Ganges and peak flooding period was from September 15 to October 15, 2003. Flood maps for different dates of 2003 and different years derived from various satellites such as MODIS, LANDSAT (*Source:* NASA supported Dartmouth Flood Observation Center, USA) and other microwave sensors were overlaid to obtain the spatio-temporal

Figure 5. Spatio-temporal dynamics of flood during the event year 2003 and for the period 1998-2003.

dynamics of flood. Vector layer of location map and digital elevation map was also overlaid over the composite flood map (Sahoo *et al.*, 2004).

Flood risk zoning is an essential tool for appropriate land use planning in flood prone areas. It aims at demarcating areas likely to be affected by floods of different magnitude, probability levels, and risks associated for living and economic activities, which in turn specifies the types of permissible developments in these zones to be undertaken. Recently Government of India has enacted a legislation to introduce crop insurance in the event of natural calamities like flood and drought. Flood risk zoning would help in a big way to provide crop insurance cover in such areas. Inputs for preparing flood risk zoning are (i) flood inundation maps with respect to floods of different magnitude, frequencies/return periods in the historic past, (ii) land use in the flood plain, (iii) existing and planned development schemes, (iv) population (v) economic activities, (vi) legislation, (vii) building by laws and (viii) insurance policies.

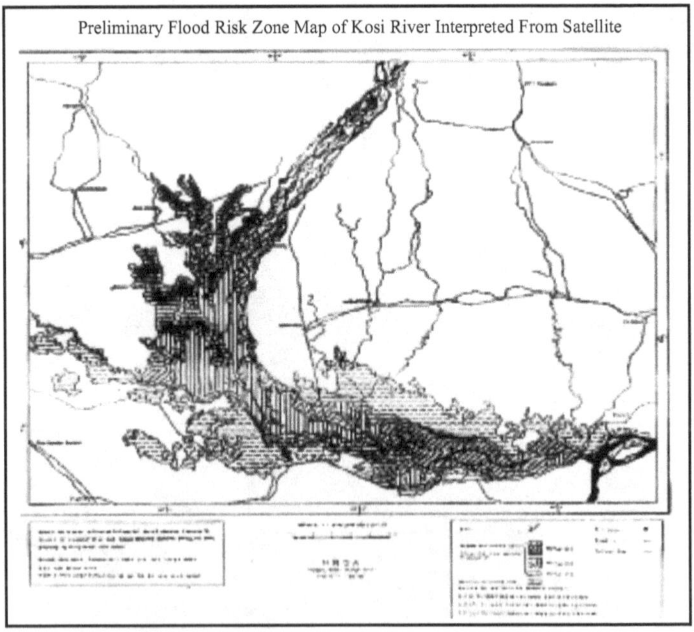

Figure 6. Flood risk zone map of Kosi River.

An attempt of National Remote Sensing Agency (NRSA), Hyderabad for flood risk zoning of Kosi river in Bihar based on limited satellite data is cited as an example in the Fig 6.

Flood forecasting models though not popular in Indian context due to non availability of spatial information required by model, is a potential approach to predict the river stage or water level, discharge and likely inundation at some future time at the downstream point of concern. The four controlling steps for real time flood forecasting using remote sensing, GIS and hydrological models are (i) real time satellite estimation of rainfall, (ii) run off estimation from the catchment using SCS TR-20 model, (iii) modeling flood flow propagation in the river using MIKE-II and (iv) modeling of likely inundation using DEM in the flood-plain.

Capturing the spatial variability of cropping pattern

Spatial variability of cropping pattern get reflected in terms of spectral variations observed on the satellite imageries, if spatial, spectral, radiometric and temporal resolutions are adequate. Capturing these variabilities enables monitoring the state and conditions of the agricultural ecosystem. Cropping pattern of an area reflects the spatial arrangement of crops and inter-crop variability that exists. Multi-date satellite data is essential to derive cropping pattern as in general there are more than one crop grown in a season with different growing calendar. Satellite remote sensing has number of advantages for monitoring cropping pattern such as (i) Multi-spectral observations which allows crop discrimination and growth assessment, (ii) Synoptic coverage at multiple spatial resolutions which leads to micro and macro level assessments over large areas, (iii) Multi-temporal, *i.e.* multiple observations in a crop season, (iv) Multiple utility *i.e.* same data can be used for many applications and (v) Full data archival and reanalysis capability *i.e.* historical primary data can be reanalyzed with new models as well as for post-harvest validation.

Cropping pattern analysis of Gangetic active flood plain of Bhagalpur district, Bihar extending from 25° 12′ 50″to 25° 24′ 31″N latitude and 86°43′37″ to 87°13′57″ E longitude was done using WiFS data of IRS-1D of the four dates (November 30, 1999, January 19, 2000, March 09, 2000 and April 03, 2000) during post-flooding period (*i.e. rabi* season) (Sahoo et al., 2006). The spatial resolution of WiFS sensor is 180 m and it has two spectral bands *i.e.* red (0.77 to 0.86 μm) and NIR (1.55 to 1.77 μm). With repeated coverage, radiometric consistency is hard to maintain between images obtained on different dates due to varying atmospheric conditions, variations in the solar illumination angles, and sensors calibration trends *etc.* Without some form of normalization, true changes in the scene between the two acquisition dates are difficult to interpret because they are masked by non-scene–dependent changes. Therefore, along with multi-temporal image registration, radiometric correction is an outstanding requirement of image pre-processing process for cropping pattern analysis (Coppin and Bauer, 1996).

The mathematical model that best describes the normalization of two images of the same area acquired at two different times involves the use of regression. The algorithm assumes that the pixels samples at day 1 are linearly related to the pixels for the same locations at day 2. This implies that the spectral reflectance properties of the samples pixels have not changed during the time interval. This sample pixels are known as pseudo-invariant features (PIF) and key to the image regression method of radiometric normalization (Schott *et al,*. 1988; Jensen, 1996; Heo and FitzHugh, 2000). PIFs are in-scene man-made elements (*e.g.,* urban areas in this study). The PIF technique is designed to permit scene-to-scene normalization by implementing a set of radiometric transformations. The techniques do not require sensor calibration or atmospheric parameters. The methodology of the PIF approach for identification of crops is given as below in Fig. 7. Ground truthing was done with GPS Leica GS-5 and Survey of India toposheets. Geometric correction was done to all the images using ground control points (GCPs).

Radiometric normalization was done based on pseudo-invariant features (PIFs) in the images, which are objects spatially well defined and spectrally and radiometrically stable.

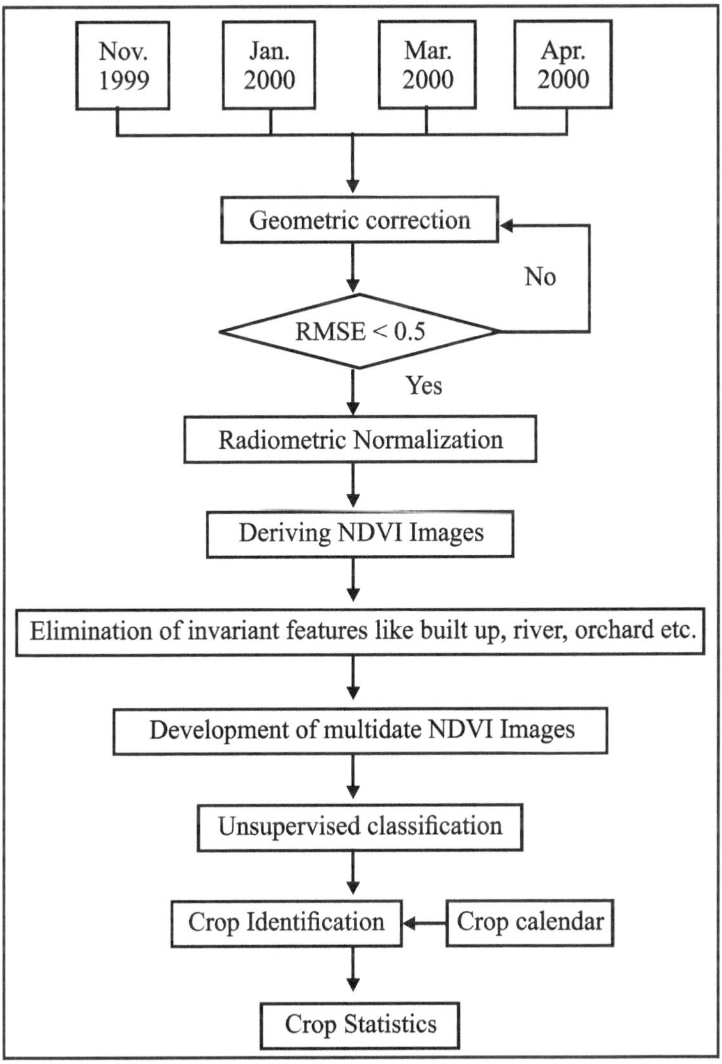

Figure 7. Methodology for capturing spatial cropping pattern distribution using temporal satellite data.

The ideal PIFs are those that meet the following criteria: (i) they are at approximately at the same elevation as the rest of the scene (for a better representation of the atmospheric conditions across the scene), (ii) they are in a relatively flat area (to minimize the effects of solar azimuth differences), and

(iii) should have a minimal amount of vegetation (as vegetation readily changes in response to seasonal changes and environmental stresses) (Eckhardt *et al.*, 1990). The assumptions considered while selecting PIFs of the study area were such as (i) the DN values of the PIF pixels involved in correction procedure are invariant during the period under study, (ii) the variation in the image digital number (DN) during the period are linear and spatially homogenous for all PIFs involved in the normalization procedure and (iii) linear effects are much greater than that of non-linear.

Image having highest dynamic DN range i.e. April, 2000 was selected as the reference (or base) image for radiometric normalization of other three images (*i.e.* Nov,1999, Jan, 2000 and Mar, 2000). The PIFs such as urban area, helipad, play ground *etc* from the study area were segmented from images of different dates. If these PIF can be isolated on day 1 and day 2, then their reflectivity distribution in each spectral channel should represent samples from the same reflectance population. These reflectance distributions should be linearly related if the samples indeed represent a statistically invariant population. For objects with the same reflectivity on both the dates, the digital counts on each day are simple linear functions of the same variables and therefore are linear functions of each other. Thus, according to Schott *et al* (1988)

$$DN_{1i} = \frac{\sigma_{1i}}{\sigma_{2i}} DN_{2i} + \left(\mu_{1i} - \frac{\sigma_{1i}}{\sigma_{2i}} \mu_{2i} \right)$$

where,

DN_{1i} = Pixel Value of day1 *

DN_{2i} = Pixel Value of day2**

σ_{1i} = Standard deviation (SD) of PIF of day1*

σ_{2i} = Standard deviation (SD)of PIF of day2**

μ_{1i} = Mean of PIF of day1*

μ_{2i} = Mean of PIF of day2**

i = band number (i=1,2,3,4)

*day1 = April or its equivalent

**day2 = November/January/March

The mean and standard deviation values of PIFs of each band of images of four dates were retrieved and given in Table 1 and using these values in above formula band-to-band radiometric normalization was done. The table also shows comparison between the mean and standard deviation values of December, January and March images with April at respective band levels. The normalized bands of images of respective dates were stacked to get the FCC for further analysis.

The Normalized Difference Vegetation Index (NDVI) a potential indicator for crop growth and vigour was used in the study and expressed as

$$NDVI = \frac{(NIR - Red)}{(NIR + Red)}$$

Where, NIR and Red are reflectances / DN values in near infrared and red bands. NDVI image was prepared from images of four dates and scaled to unsigned 8bit (0-255) values. The spectral profile of different crop classes could be analyzed using temporal NDVI images. All four NDVI images were stacked together as multidate NDVI image (Fig. 8) having November, January, March and April NDVIs as band 1, 2, 3 and 4 respectively. The invariant features like water, barren lands, orchards etc were found to have NDVI spectral profile as almost straight line and have been removed from the image. The remaining features in multidate NDVI image were all agricultural crops, which is of our interest to identify them. The multidate stacked NDVI image having only cropped area was clustered into 10 classes and temporal spectral profile of NDVI was plotted for each class as shown in Fig 9.

Ground truth and other crop information collected were used in identifying the NDVI spectral pattern and assigning it to a particular crop class. Since the temporal NDVI spectral pattern, which is representative of the crop dynamics, is specific to a crop, it facilitates in identifying the crop. Based on crop calendar and ground truth (GPS data), the temporal NDVI

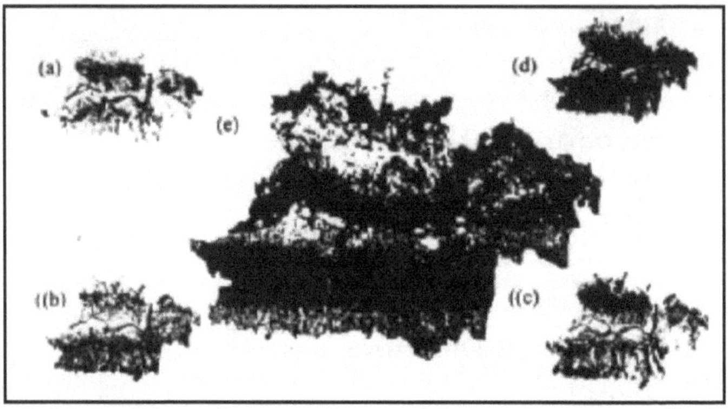

Figure 8. Methodology for capturing spatial cropping pattern distribution using temporal satellite data.

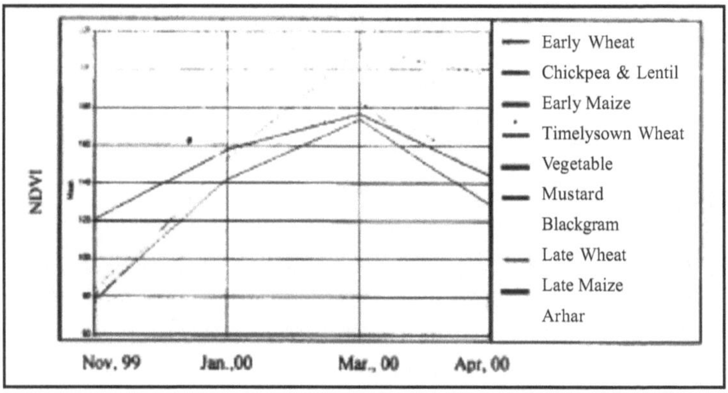

Figure 9. Temporal NDVI profile of different crops.

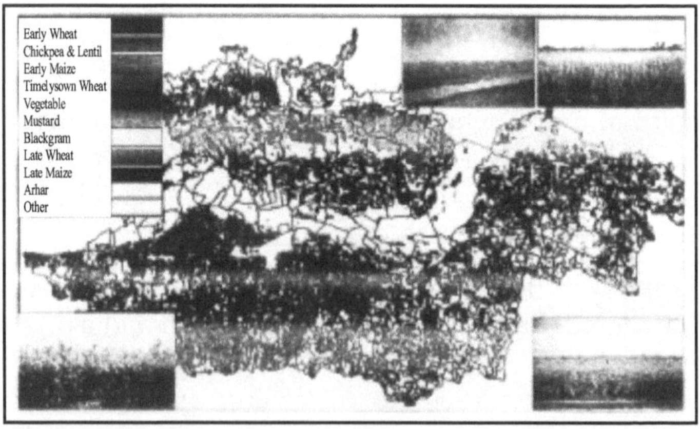

Figure 10. Spatial distribution of crops grown (*rabi*, 1999-2000) overlaid with village boundaries of Bhagalpur district, Bihar.

profile of each class was identified as different crop and their spatial distribution is given in Fig 10. Even if the crops like wheat and maize having sown early and late could have been identified from significant difference in its NDVI profile pattern. The growing of vegetables has good association to riverbanks and pulses like lentil, pea with wet land locally known as *tal* lands and black gram active flood plain (particularly to diara lands). The specific association of crop to growing areas helped to great extent in identifying it. The other class in figure includes invariant features such as river, sand beds, barren land, built up, water bodies, orchards *etc.* The area statistics of different crops was computed and given in Table 2.

It was found that out of about 2.27 lakh ha of study area, various agricultural crops cover 59 per cent. The wheat crop was found to be dominating one (20%) followed by maize (12 %) and lentil and chick pea (10%). In wheat growing areas, about 2, 12 and 6%area was occupied by early, timely and late sown wheat, respectively. Similarly, in maize crop, 8 and 4% was in early and late sown categories respectively. Winter maize and wheat is grown majorly in flood plains close to river bank (diara land), pulses are common in low lying areas called *tal* lands. Arhar crop sown during May/June if not damaged by flood remains in the field upto late March /early April. Other 41% includes built up area, orchards/plantation areas, barren lands, river, water bodies etc. Digitized map of Bhagalpur with village boundaries was also overlaid to retrieve the information about late planted crops and fallow lands for possible interventions.

Addressing the seasonal dynamism of agricultural system

Dynamism of crops over an agricultural calendar can be captured from 10 days maximum value composite satellite data products like NDVI images of SPOT-4 vegetation sensor. NDVI data from a satellite sensor is primarily related to vegetation changes and follows annual cycle of growth and decline. Thirty six NDVI images of 10 days composite for the agricultural

Table 1. Mean and Standard Deviation values of PIFs of four dates before and after normalization.

Date_band	Mean	SD	Mean	SD
	Before Normalization		After Normalization	
Nov_1	57.92	5.57	102.87	3.28
Nov_2	59.07	6.99	113.63	2.71
Jan_1	68.43	5.78	104.56	3.40
Jan_2	59.81	6.85	114.50	3.16
Mar_1	79.61	10.15	110.56	11.94
Mar_2	86.49	9.64	120.61	7.29
Apr_1	100.06	14.94	100.06	14.94
Apr_2	106.78	12.43	106.78	12.43

Table 2. Area statistics of crops of the study area.

Crop Classes	No of Pixels	Area (ha)	Area (%)
Early sown Wheat	972	3620.6	2
Timely sown Wheat	7411	27605.2	12
Late sown Wheat	3702	13789.6	6
Chick pea and lentil	5810	21641.7	10
Early Maize	5034	18751.1	8
Late Maize	2162	8053.2	4
Vegetable	2700	10057.2	4
Mustard	3546	13208.5	6
Black gram	2686	10005.1	4
Arhar	2101	7826.0	3
Total cropped area	36124	134558.3	59
Others	24787	92329.1	41
Grand Total	60911	226887.4	100

calendar year June 1999-May 2000 were stacked together to derive temporal NDVI profile (Sahoo *et al.*, 2004). Clouds and poor atmospheric conditions usually depress NDVI values causing sudden drops in NDVI which are not compatible with the gradual process of vegetation growth and is considered as noise. This noise was removed using Fourier Adjusted Sun Zenith Angle Corrected Interpolated and Reconstructed (FASIR) Technique. Corrected image was used to retrieve smooth temporal NDVI profile (Sahoo et al., 2004), were analyzed using ground information and whole area was classified into four classes of depending on cropped and fallow lands during *kharif* and *rabi* seasons (Fig .11).

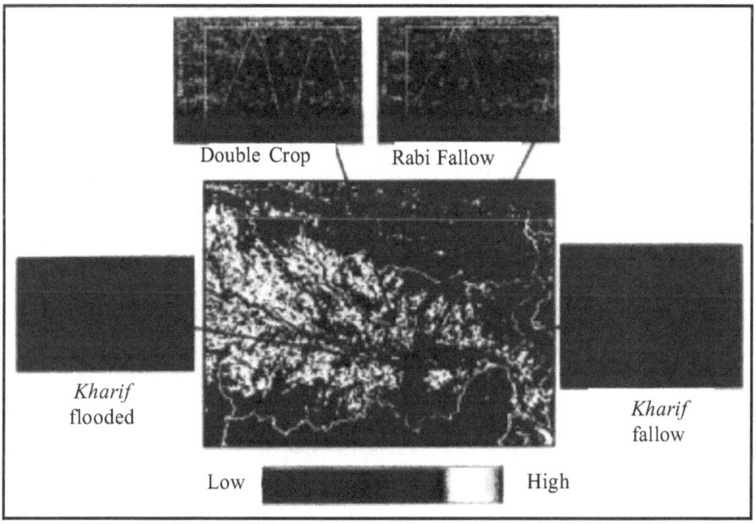

Figure 11. Stacked SPOT NDVI image showing its temporal profile of a year for *kharif* and *rabi* fallow, *kharif* flooded and doubled cropped area.

Targeting for possible interventions to enhance the productivity

Satellite remote sensing along with Geographic Information System helps in retrieving and integrating various information to make possible decision on intervention for enhancing productivity at spatio-temporal scale. For example spatio-temporal pattern of flood and crops in flood plain agro-ecosystem indirectly gives insight for targeting technologies. As discussed in the previous sections, remote sensing could be used to know potential flooding areas, its recession pattern and land availability for cultivation which decides the kind of cropping pattern practiced in the region.

Many lands having high production potential remains fallow due to late harvesting of *kharif* crop and longer turn around time between two crops in the system. In upland flood plain areas where risk of flood is minimal and land is available through out the year, coarse or fine rice crop is taken during *kharif* season. Timely harvest of coarse rice leads to timely sowing of wheat. But the late harvest of fine rice during December results in late planting of wheat, which is not profitable for farmer. Hence, they either prefer to keep the land fallow or grow vegetable crops in some patches. Land remaining fallow after harvest of rice in *kharif* is called rice-fallow. Inventory on rice fallow made by NRSA and ICRISAT through IRS WiFS data analysis during 1999-2000 revealed that India alone has 11.65 m.ha, which is 29 % of rice area. Nearly 82 % rice fallows are located in the most flood prone states. GIS analysis of these fallow lands has indicated that they represent diverse soil types and climatic conditions, thus crop diversification through growing varieties of warm and cool season legumes can be best suitable option having multi-facetd impact on the economy. In such areas also mechanization of crop harvesting and use of different Resource Conserving Technologies (RCTs) for timely planting would bring the fallow lands under cultivation and can significantly increase the productivity of the system. Resource Conserving Technologies include zero, minimum and reduced tillages, surface seeding etc which not only save land preparation time but also increase water and nutrient use efficiencies

Again, generally flooding period in Eastern Gangetic plains (namely Bhagalpur district of Bihar) ranges from 3rd week of July to 4th week of September. In these areas, sowing schedules of crops (mainly wheat and winter maize) depends exclusively on the recession pattern of flood, which decides the availability of land for cultivation. Early recession of flood allows farmers to take black gram followed by either late wheat or winter maize and summer maize. Black gram is a bonus crop as it is broadcasted in the fields to grow on residual soil moisture, there by it does not require any land preparation and inputs except the seeds. But this crop is harvested in December which delays planting of succeeding wheat and winter maize, consequently affecting its yield and also indirectly secure harvest of the summer maize before onset of next flood. Traditionally, in these areas, wheat crop is sown by broadcasting method after five ploughings at the seed rate of 250 kg ha^{-1} and basal fertilizer is broadcasted on the surface. Due to ploughing operations, late planting of wheat crop results in poor yields due to reduction in growing period of the crop. To avoid late planting and reduce the seed rate and for proper placement of basal fertilizers in the root zone, seed cum fertilizer drill (*i.e* one of the RCTs) could be a potential tool. From the exploratory trial in farmers fields in diara lands of Sabour block, Bhagalpur under aegis of Rice-Wheat Consortium, CIMMYT, New Delhi, it was found that Seed-cum-fertilizer drill could reduce commonly practiced five ploughings to two which were good enough for proper placement of seed and fertilizers and reduced the seed rate by about 150 kg ha^{-1}. This also resulted in producing long and healthy ears and hence the higher yields compared to traditional system (Sahoo *et al.,* 2003).

Conclusions

Agriculture systems to survive, have to be ecologically sound and more profitable by virtue of enhanced productivity at lower cost of cultivation. It needs to be therefore transition to a more information and knowledge based so that it could address not only the issue of poverty alleviation but also serve as an engine to the accelerated growth in the region. Remote sensing technology found to be most suited to make agriculture to be information and knowledge and to monitor also associated yield reducing environmental events so that timely generated data and information extracted can further be used for strategic improvement of agricultural productivity in a stress environment. This can be concluded that satellite remote sensing could capture spatio-temporal dynamics of flood and crop, delineating the areas where lands remain fallow, covered by late wheat and maize crops and also temporal availability of land for cultivation which could help in targeting possible interventions to enhance the productivity. Future use of remote sensing, GIS and GPS technologies in flood plain agro-ecosystem is mainly for sustainable agricultural development of the region with reducing impact of disaster like flood on it.

Acknowledgement

The study was carried out under a project funded by National Agricultural Technology Project, ICAR and is duly acknowledged. The authors also duly acknowledge RRSSC, Kharagpur and DMS, ISRO, Bangalore and Water Resource Division, NRSA for sharing their valuable information for this study.

References

Anonymous, 1974-1981. Report of the Study Team on Diara areas of Uttar Pradesh and Bihar. Indian Council of Agricultural Research.

Center for Science and Environment (CSE), 1991. Floods, Flood Plains, and Environmental Myths', CSE, New Delhi.

Centre for Science and Environment (CSE), 1987. The Wrath of Nature: The impact of environment destruction on floods and drought. CSE, New Delhi.

Central Water Commission, Statewise data on flood affected areas and damages. 1953 - 1984. New Delhi, mimeo.

Coppin, P.R. and Bauer, M.E. 1996. Processing of multi-temporal Landsat TM imagery to optimize extraction of forest cover change features. *IEEE Trans. Geosci. Remote Sens*. 32(4): 918-927.

Eckhardt, D.W., Verdin, J.P. and Lyford, G.R. 1990. Automated update of an irrigated lands GIS using SPOT HRV imagery, *Photogrammetric Engineering and Remote Sensing* 56(11): 1515-1522.

Gangwar, B. 1999. Technological advances for increasing crop productivity in Diara areas. *Agrl. Situation in India*. 615-620.

Heo, J. & FitzHugh W. T. 2000. A standardized radiometric normalization method for change detection using remotely sensed imagery, *Photogrammetric Engineering and Remote Sensing*, 66(2): 173-181.

Hofer, T. 1993. Himalayan deforestation, changing river discharge and increasing floods: Myth or Reality? *Mountain Research and Development*, 13(3):213-233.

Jensen, J.R. 1996. Image Preprocessing: Radiometric and Geometric Correction. In : Introductory Digital Image Processing. Prentice Hall Publisher (Second Edition): 107-135.

Rashtriya Barh Ayog 1980. Report. Vol.1, Ministry of Energy and Irrigation. New Delhi.

Sahoo, R.N., Tomar, R.K., Sehgal, V.K., Rao, C.S., Nirupa Charchi., Abrol, I.P., Wadhwani, M.K. and Gupta, R.K.(2004). Geo-informatics for enhancing crop productivity in Eastern flood plains-A case study. Proc. Natl. Symp. on *"Geo-informatics Applications for Sustainable Development"* Feb. 17-19, IARI, New Delhi : 183-189.

Sahoo, R.N., Tomar, R.K., Abrol, I.P., Krishnamurty, V.V.R.S., Wadhwani, M.K.,and Tiwari, M.K.(2002): Drivers of land use dynamics in Gangetic plains: Application of Remote sensing and GIS analysis. Proc. LUCC-ISPRS workshop *"LUCC Contribution to Asian Environmental Problems,* Dec. 2, NRSA, Hyderabad, pp. 20-21.

Sahoo, R.N., Tomar, R.K., Rao, C.S., Abrol, I.P., Wadhawani, M.K. (2003). Potential of RCTs in Eastern Gangetic Flood Plain: A Remote Sensing Analysis, In Rice-Wheat Information Sheet, Rice Wheat Consortium for the Indo-Gangetic Plains (RWC), CIMMYT, New Delhi, (Rice Wheat Information Sheet, RWIS), 47:8-9.

Sahoo, R.N., Tomar, R.K., Rao, C.S., Sehgal, V.K., Charchi, Nirupa, Abrol, I.P., Tiwari, M.K. and Wadhawani, M.K. (2006). Radiometric Scene Correction of Temporal Multispectral Satellite Data for Discrimination of Crops, *Indian Journal of Radio and Space Physics.* (35: 116-121).

Schott, J.R. Salvaggio C. and Volchok, W.J. (1988) Radiometric scene normalization using pseudoinvariant features, *Remote Sens. Environ. 26: 1-16.*

Venkatachary, K.V, Bandyopadhaya, K., Bhanumurty, V, Rao, G.S., Sudhakar, S. Pal, D.K. Das, R.K., Sharma, Utpal, Manikam, B, Meena Rani, H.C. and Srivastava, S.K. 2001 Defining a space based disaster management system for floods: A case study for damage assessement due to 1998 Brahmaputra Floods, *Current Science* 80 (30) pp. 369-377.

❑❑❑

GEOINFORMATICS APPLICATIONS IN AGRICULTURE, 2007
Eds. : A.K. Singh and U.K. Chopra; pp. 271-291
New India Publishing Agency, New Delhi (India)
E-mail : newindiapublishingagency@gmail.com
Web: www.bookfactoryindia.com

10
Groundwater Modelling Using GIS

M.S. Mane, D.K.Singh, A.K.Singh & A.K.Bhattacharya

Key words : Groundwater Modelling, GIS, PMWIN, ArcInfo Interface Con2grid, Interface Programme, ArcMacro Language.

Introduction

Groundwater is an important source of irrigation in India and has played a major role in increasing the food production. About 50 per cent of total irrigated area in the country is dependent on groundwater (Central Water Commission (CWC), 2000). Productivity of groundwater is normally more than the surface water due to the fact that it is available at the point of use, requires minimum conveyance infrastructure, is available on demand and maximizes the application efficiency. However, in past few years, excessive and unplanned groundwater pumping in certain parts of the country has resulted in depletion of groundwater resources. Water table in these region is going down day by day and making it beyond the reach of poor farmers. Therefore, keeping the water table in the desirable range is a major concern. This can be achieved by adopting proper pumping rate and pumping schedule. Groundwater modelling has successfully been used for deciding the pumping rate and pumping schedule. Groundwater modelling refers to simulation of groundwater behaviour under varying pumping and recharge conditions. A groundwater model is simply a set of equations, which describe the physical processes active in the aquifer subjected to certain assumptions. While the model itself obviously lacks the detailed reality of the groundwater systems, the behaviour of the validated model approximates that of the aquifer.

Groundwater modeling often involves handling large input and output data sets. Geographic Information System (GIS) provides an integrated platform to manage, analyze and display different data sets and can greatly facilitate modeling efforts in data compilation, model calibration and display of model parameters and results.

Presently, in India, GIS is being used mostly for groundwater mapping and creation of information base. Modelling capability of GIS has not been used much in India but has been used extensively in other countries.

GIS provides the unique advantage of integrating the groundwater simulation model and maps which is also referred as map based groundwater simulation. Another advantage of map based simulation is that results of groundwater simulation can also be presented in global reference coordinate system, which makes it easier for the model result to be interpreted and used than when the groundwater model is constructed in model coordinate system whose exact geographic location has not been precisely defined.

Flow of groundwater in an aquifer is governed by two or three dimensional partial differential equations subjected to different boundary conditions. The availability of high speed computers and concurrent development of efficient numerical methods for the solution of partial differential equations of groundwater flow have enabled the development of complex groundwater flow models. The partial differential equations are generally solved either by Finite Difference Method (FDM) or Finite Element Method (FEM). The basic principle behind both these methods is to replace the true solution by an approximate one constructed from a suitable set of basic functions. FDM is a technique for solving the governing partial differential equation in which continuous flow domain is replaced by a finite number of regular spaced mesh- or grid-points called as nodes representing volume-averaged sub-domain properties. Soultion is obtained by approximating the derivatives of the partial differential equation for each of these points using finite differences. Extensive reviews of groundwater models are available in literature (Prickett, 1976; Pinder and Gray, 1977; Rushton, 1988; Storetz and Bradburry, 1981; Krabbenhoft and Babiarz, 1992). The resulting set of linear or nonlinear algebraic equations is solved using direct or iterative matrix solving techniques. FEM is similar to Finite Difference Method with the exception that the mesh may consist of regular or irregular-spaced grid points which may have irregular shapes; and the PDE is approximated using the method of weighted residuals to obtain a set of algebraic equations. These algebraic equations are solved using direct or iterative matrix solving techniques.

The commonly used numerical algorithams are Iterativge Alternating Direction Implicit Method (IADI) (Pricket and Lonnguist, 1971) and standard backward difference scheme (Rushton and Redshaw, 1979). McDonald and Harbaugh (1988) developed a modular three dimensional finite difference groundwater flow model called 'MODFLOW' which is extensively used for simulation of groundwater flow. MODFLOW uses the following equations for groundwater flow simulation:

Continuity Equation

The method of finite differences with block centered approach is adopted and the resulting equations are solved using the continuity equation (Eq. 1) with an assumption that the density is constant. In the block centered approach, the nodes are in the center of the cells and parameters are taken as constant within a cell.

$$\sum Q_i = SS^{\Delta h} - \Delta V \qquad \text{... (1)}$$

where,

Q_i = flow rate into the cell (L^3/T),

SS = a term equivalent to specific storage (1/L),

V = volume of the cell (L^3), and

Darcy's Law

Darcy's equation to describe flow into cell (i, j, k) from cell (i, j-1, k).

$$q_{i,j-1/2,k} = KR_{i,j-1/2,k}\Delta c_i \Delta v_k \frac{h_{i,j-1,k} - h_{i,j,k}}{\Delta r_{j-1/2}} \quad \text{... (2)}$$

Δr_j = Dimensions of the cell along column direction

Δc_i = Dimensions of the cell along column direction

Δv_k = Dimensions of the cell along column direction

$KR_{i,j-1/2,k}$ = Hydraulic conductivity in row i and layer k between nodes i,j-1,k and i,j,k

Finite Difference Model

The governing equation in the finite difference model of groundwater flow is given by:

$$\frac{\partial}{\partial x}\left(K_{xx}\frac{\partial h}{\partial x}\right)+\frac{\partial}{\partial y}\left(K_{yy}\frac{\partial h}{\partial y}\right)+\frac{\partial}{\partial z}\left(K_{zz}\frac{\partial h}{\partial z}\right)-W = S_s\frac{\partial h}{\partial t} \quad \text{...(3)}$$

Where, Kxx, Kyy, and Kzz are values of hydraulic conductivity along the x, y, and z coordinate axes, which are assumed to be parallel to the major axes of hydraulic conductivity (L/T);

h= potentiometric head (L);

W= volumetric flux per unit volume representing sources and/or sinks of water, with $W<0.0$ for flow out of the groundwater system & $W>0.0$ for flow in ground water system (T^{-1});

Ss= specific storage of the porous material (L^{-1}); and

t= time (T).

Watkins *et al.*, (1996) used MODFLOW that employs rectangular discretization in which the parameter values were entered and the hydraulic heads were computed for each cell. They assumed that the flow in the aquifer was horizontal. Stacking of grid layers was done to specify the interaction between them, which often represent the three dimensional finite difference models.

Johnson *et al.*, (1998) stated that the numerical models such as MODFLOW provide an excellent tool to assist in the water resource decision process. A simplified representation of the system may in some cases, be obtained by the development of response function from numerical models. It was also reported that MODFLOW could produce predictive simulations of cause and effect when properly calibrated and applied to an area.

Tsou and Whittemore (2001) developed an ArcView interface for the groundwater flow and trasport models MODFLOW and MT3D. The extension included pre processing of spatially distributed data for model input and post processing of model output. Models could be automatically calibrated through the ArcView interface by external linking to such programs as PEST. The efficient pre and post processing capabilities and calibration link were demonstrated for groundwater modeling in southwest Kansas.

An ArcInfo interface *Con2grid* for groundwater flow model Processing MODFLOW for Windows (PMWIN) was built as an extension for facilitating modeling (Mane, 2004; Mane *et al*, 2004). The extension includes pre processing of spatially distributed (point, line and polygon) data for model input. With the help of ArcInfo interface *Con2grid* any coverage of spatially distributed data can be converted into ASCII files of user defined grid dimensions which can be used as input to the model. The efficient pre processing capabilities were demonstrated for groundwater modeling in Western Indo-Gangatic Plains of Utter Pradesh in India.

GIS has been coupled to simple functional models of recharge to map a regional assessment of relative potential recharge (Boniol *et al.*, 1993; Fayer *et al.*, 1996; Rogowski, 1996). Maidment (1993) presented an excellent review of GIS applications in hydrologic modeling. GIS has been successfully interfaced with groundwater flow and solute transport models. Harries *et al.*, (1993) and Wilson *et al.*, (1996) reported integration of the ArcInfo GIS with groundwater flow and contaminant transport models. Extensive reviews on coupling of GIS to simulate flow of water and non-point source pollutants in aquifers were covered by Corwin and Wagenet (1996) and Goodchild *et al.* (1993).

Map based Groundwater Model

Ye *et al.,* (1996) of University of Texas developed a map based groundwater simulation model utilizing the concept of object oriented programming and GIS. The model is constructed using the relations between the cells (Polygons) and their boundaries (Arcs) kept in ARC/INFO polygon and arc coverages (Fig.1 to 3).

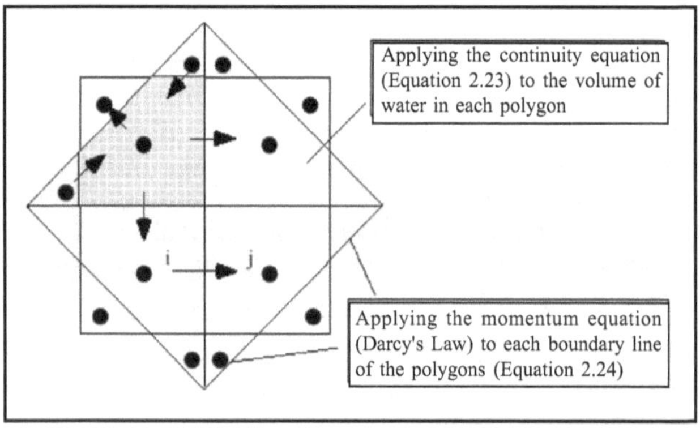

Figure 1. The conceptual design of Map based Groundwater Model
Source: http:// www.ce.utexas.edu/prof/maidment/gishydro/docs/reports/ye/ Sect4.pdf

To work within the spatial databases, model first applies the Darcy's law to each boundary line object to calculate the volumes of flow defined on the boundary line objects, together with recharge, pumping, and spring flow time series defined on the polygons. The continuity equation is applied to the polygons to calculate the water levels at the end of time step. This process is repeated until the final time step is reached.

Modules of the groundwater simulation model developed by University of Texas are listed:

- *Module I, Pre-procesor:* Construction of the groundwater modelling objects (Cell polygons and cell boundary line object) from GIS maps
- *Module II, Processor:* Using object maps created by pre-processor , processor simulates groundwater flow on the line objects based on the momentum equation and computes the water levels on the polygon objects based on the continuity equation
- *Module III, Post Processor:* Post processor analyses and displays the results. Post processor is also used for model modification and data management.

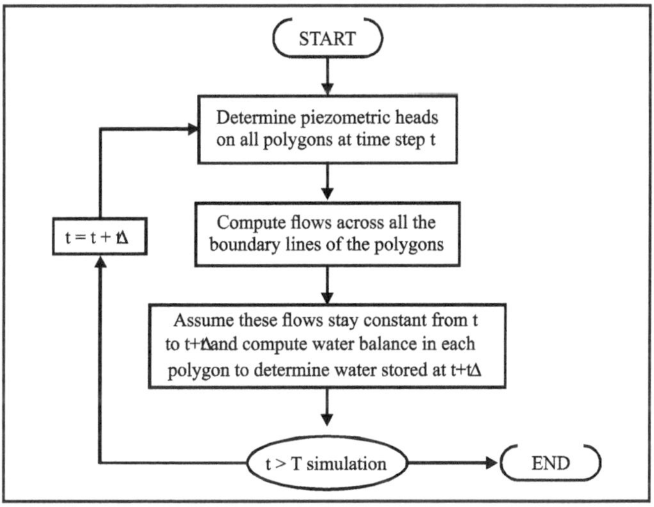

Figure 2. Programme flow chart of the Groundwater Simulation Model *Source:* http:// www.ce.utexas.edu/prof/maidment/gishydro/docs/reports/ye/ Sect4.pdf

```
For each timestep in 1 Number of Time Steps
    For each line in the Arc Coverage
        {Compute the mass flow Across the line using Darcy's Law}
    End
    For each cell in The Polygon Coverage
        {Compute the new water level of the cell using the continuity equation}
    End
End
```

Figure 3. Three loops in Groundwater simulation algorithm
Source: http:// www.ce.utexas.edu/prof/maidment/gishydro/docs/reports/ye/ Sect4.pdf

The main programme of this model consists of three loops:

- *The time loop:* Simulates groundwater flow situations for each time step. The time loop becomes iterative process under steady state.
- *Line loop:* For each time step on the time loop, the time loop is first carried out to compute water flow across each time object based on the piezometric heads computed at the last time steps using the Darcy's law
- *Polygon loop:* Polygon loop is carried out to calculate the water level changes in each cell polygon object based on the continuity equation.

This model was applied to solve the problems of groundwater flow in phreatic aquifer with dual river system to see if model results match the theoretical solution.

ArcInfo Interface Con2grid for Groundwater Model-Processing MODFLOW for Windows (PMWIN)

An ArcInfo interface *Con2grid* was used to predict the spatial changes in water table in the command area of the Dadupur distributory under the Upper Ganga canal system (Mane, 2004; Mane et al., 2004).

The study area is a part of Bulandshahar district of Western Utter Pradesh (Fig. 4). The area is bounded by the Upper Ganga canal in the East, river Karwan in the West and Southern administrative boundary of Khurja block in the South. It has an area of 293 sq. km, lying between $28^0\ 0'$ to $28^0\ 25'$ N latitude and $77^0\ 25'$ to $78^0\ 0'$ E longitude and depicted on Survey of India's toposheet Nos. 53 H-10, 11, 12, 14, 15 and 16 of 1:50,000 scale. It covers parts of three revenue blocks namely Bulandshahar, Sikandarabad and Khurja. In this study, an interface programme was developed to link the processing MODFLOW with GIS to simulate the water table behaviour in the command of Dadupur distributary under Upper Ganga Canal System. ArcInfo 8.0.2 was used to prepare the thematic maps, develop Digital Elevation Model (DEM) prepare the input files for model and display the results.

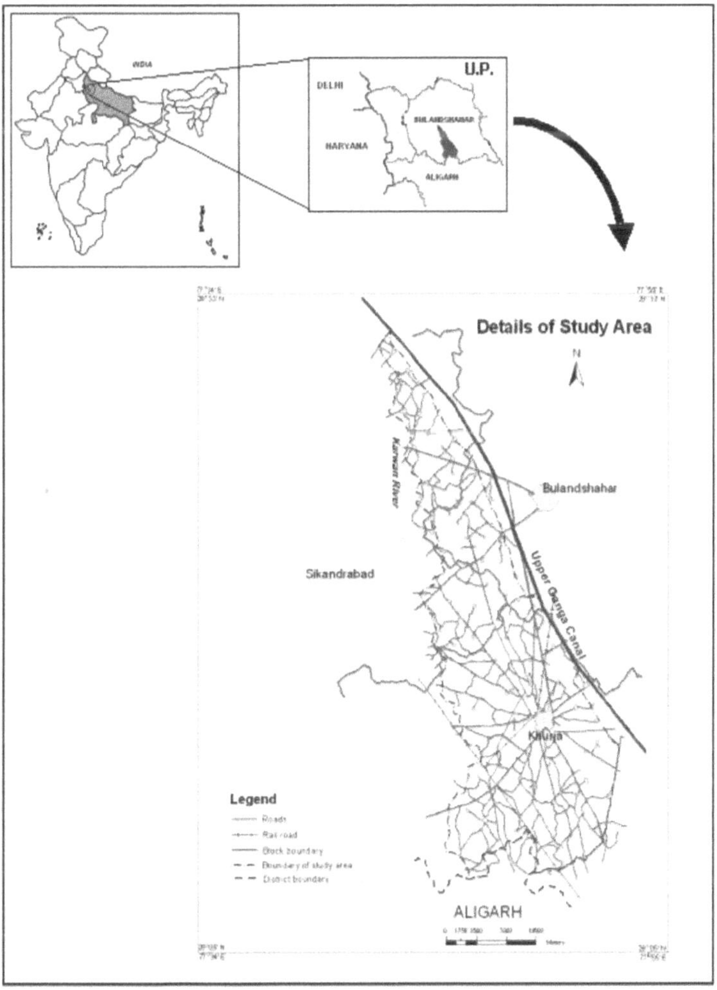

Figure 4. Location map of study area

The climate of the study area is semi-arid, with a hot summer during March to June, cold winter during November to February and a rainy season during July to October. The annual mean maximum and minimum temperature are 44.9^0 C and 3.1^0 C, mean annual rainfall is 776 mm and the mean annual potential evapotranspiration (PET) is 1578 mm.

Geomorphologically and geologically, the district is divided into two broad units as younger and older alluvium. The elevation varies from 207.4 m above MSL at the North to 193.8 m at the South boundary of the district. The general slope is above 0.29 m/km from North to South. The top 3 m of the soil is clay loam, ranging in gradation from pure sand to stiff clay, respectively known as Bhur and Matiar locally. Alluvial soils occurring in the flood plains of the yield good crop. The soil rivers map was obtained from the National Bureau of Soil Survey and Land Use Planning (NBSS & LUP) and was digitized using the ArcInfo GIS to prepare the digital soil map as shown in the Fig 5.

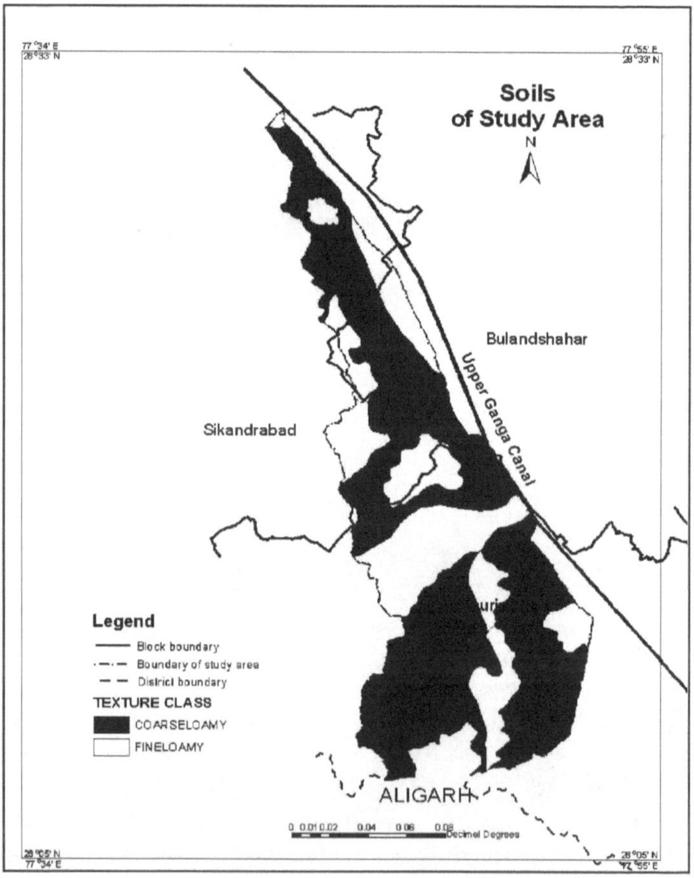

Figure 5. Soil map of the study area

The study area is in the command of the Left and Right Dadupur distributaries and their sub-branches such as Hasangardh and Mundakhera with seven minors. The distributaries and minors running through the study area are shown in Fig. 6.

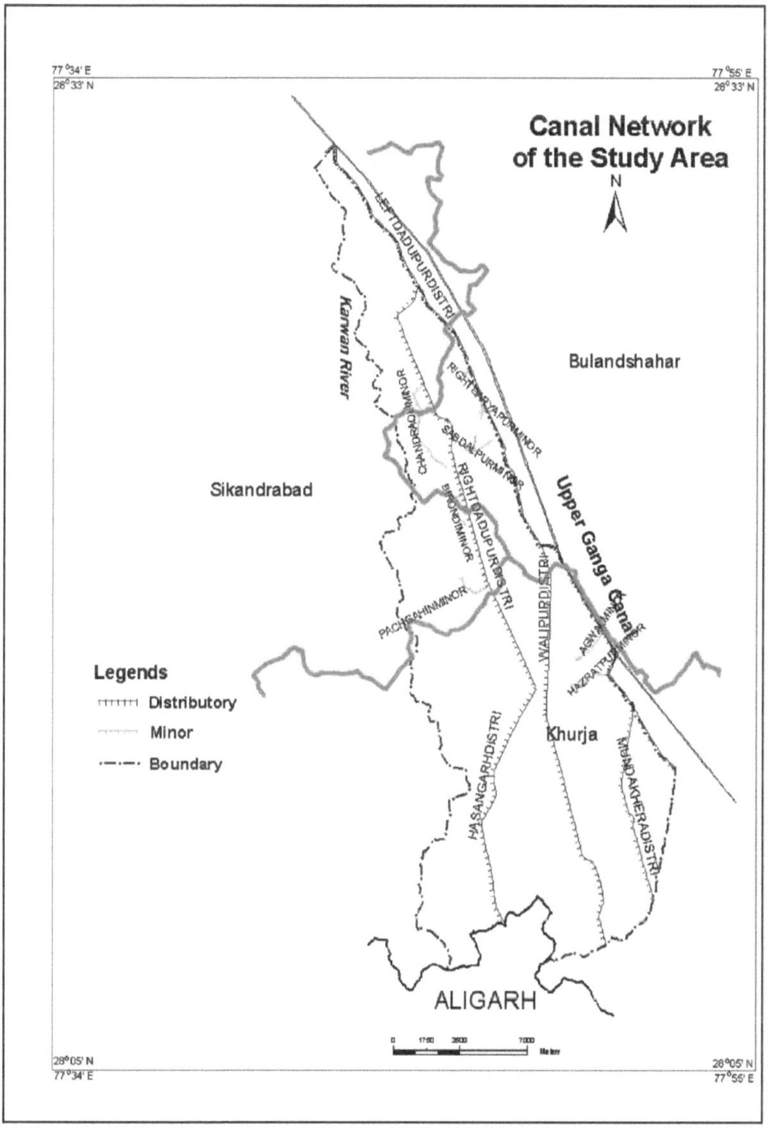

Figure 6. Map showing the canal network in the study area

To predict the spatial changes in water table, an attempt was made to link GIS with groundwater model PMWIN. The model requires various inputs like surface and bottom elevation, hydraulic conductivity, effective porosity etc. in ASC II format. As a requirement, study area need to be discretised into grids of appropriate size.

The digital elevation models of surface and bottom elevation and water table depths in different seasons were prepared using iterative finite difference interpolation technique. The output of PMWIN, which gives the grid wise water table depths, were displayed using GIS capabilities, which made the interpretation of simulation results more easy.

Various thematic maps like soil map, irrigation map etc. were prepared using ARCInfo GIS. Projected maps of various themes were used to derive information on area, canal length, land elevations, land use, soil type etc. The model needed inputs in grid format. The required grids of 1 km x 1 km size were prepared using GIS. The grid coverage is shown in Fig. 7.

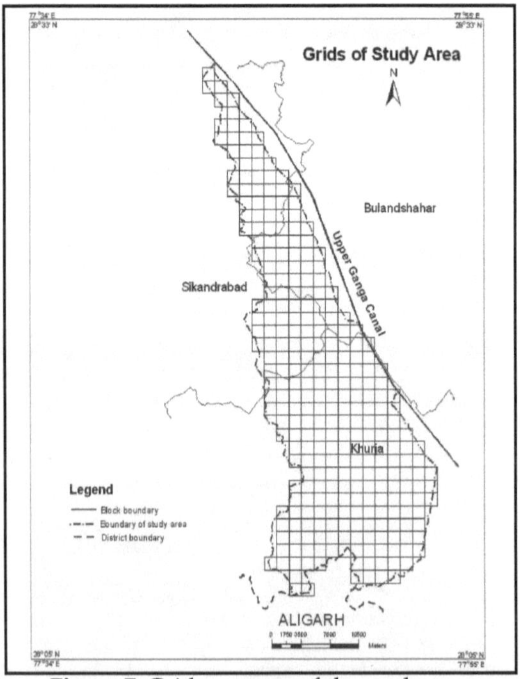

Figure 7. Grid coverage of the study area

DEM of surface and bottom elevations were prepared using the information on depth of aquifer and elevation of various pints. These are are presented in Figures 8 & 9.

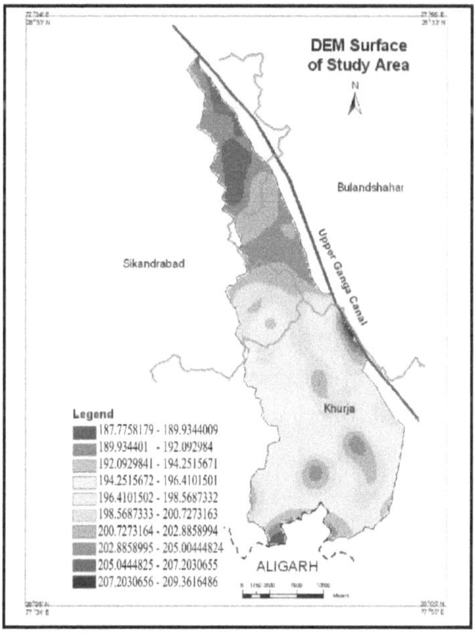

Figure 8. DEM of surface elevation

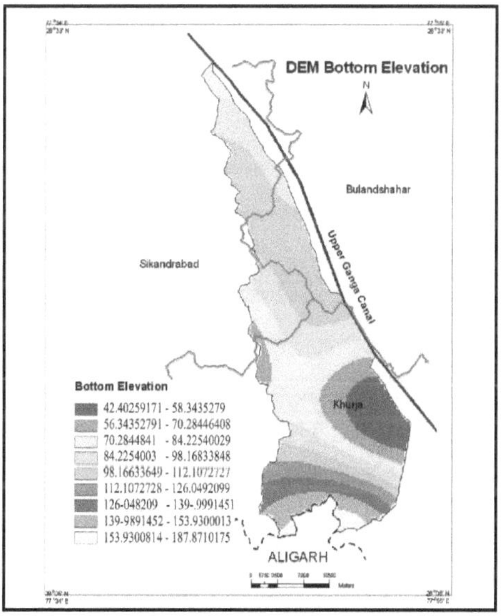

Figure 9. DEM of bottom elevation

Interface programme Con2grid was developed in Arc Macro Language, which converts the point, line or polygon coverage in grid format of user defined size and then in ASCII files and prepares the output in ASCII files. These files were directly used as input files by model after changing the header file information. The output files generated by model can be imported in ArcInfo GIS for preparing the DEM of predicted water tables. The flow chart of the Con2grid is presented in Fig. 10.

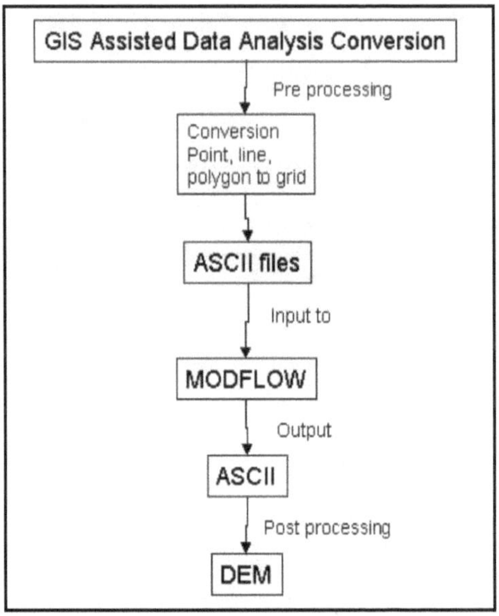

Figure 10. System architecture of *Con2grid*

A typical menu of *Con2grid* model is given in the Fig. 11

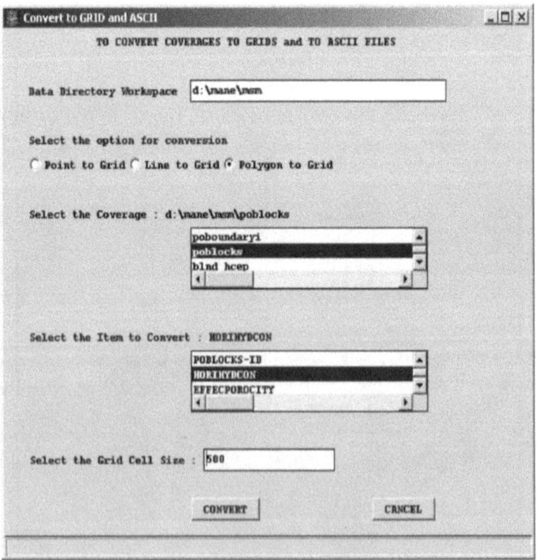

Figure 11. A typical menu of *Con2grid* model

The various attributes were attached to the polygons in the respective coverages. These coverages were the inputs to *Con2grid*. The output of *Con2grid* is shown in Fig. 12.

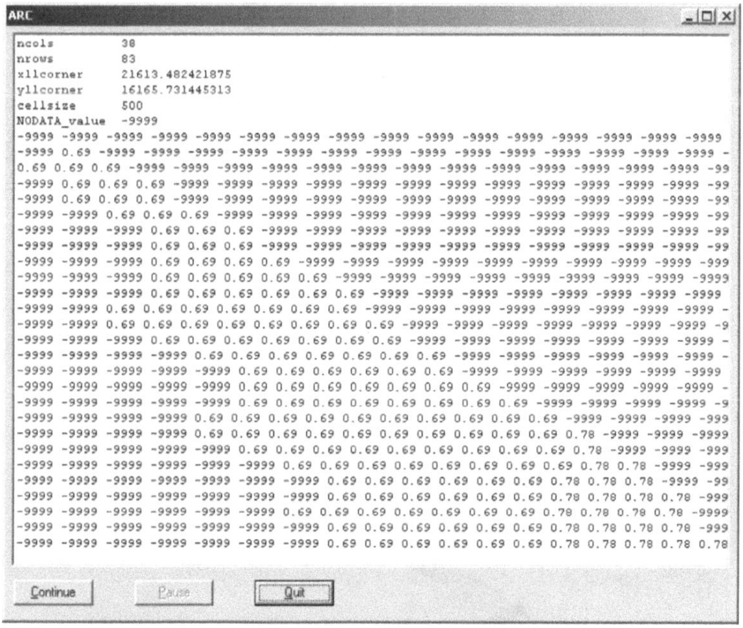

Figure 12. Output of the con2grid model

Calibration and validation of PMWIN

PMWIN was calibrated for transient simulation. Simulated and measured values of water levels were compared and aquifer parameters were adjusted to improve the model predictability. The calibration was done for the pre to post monsoon water levels of 1994. Thereafter, model was validated for post monsoon (1994) to pre monsoon (1995) water levels. The non-availability of continuous water table data was a constraint in calibration of model for longer period. Comparison of observed and predicted values of water levels for each grid are shown in Figures 13 (a & b). The scatter plot and best fit curve (R^2=0.967) showed a close agreement between observed and predicted water table depths. Comparison of predicted and observed values for these wells were also in close agreements. Result of the model validation is presented in Table 1.

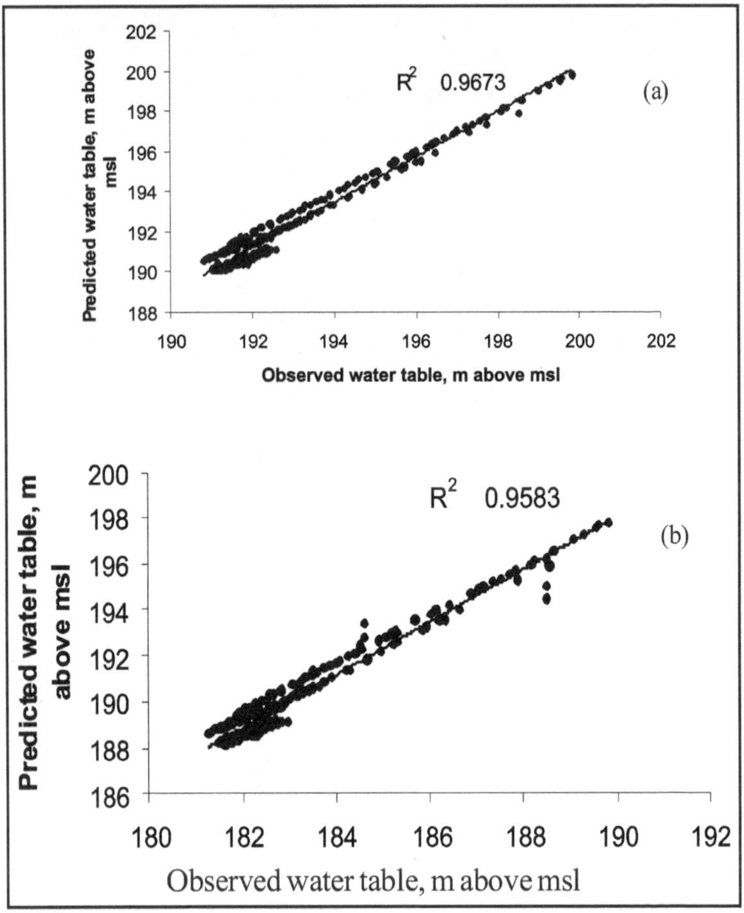

Figure 13. Result of model calibration (a) and validation (b) and

Table 1. The observed and predicted water table depths in observation wells

Sr. No.	Observation well no.	Days after initial levels	Observed water table depth, m above msl	
			Observed	Predicted
1	Located at North side	180	198.468	198.677
2	Located at South East side	180	191.850	191.073
3	Located at Southern West side	180	191.670	191.376

DEM of predicted water table depths can be generated using *Con2grid*. One such DEM is shown in Fig. 14.

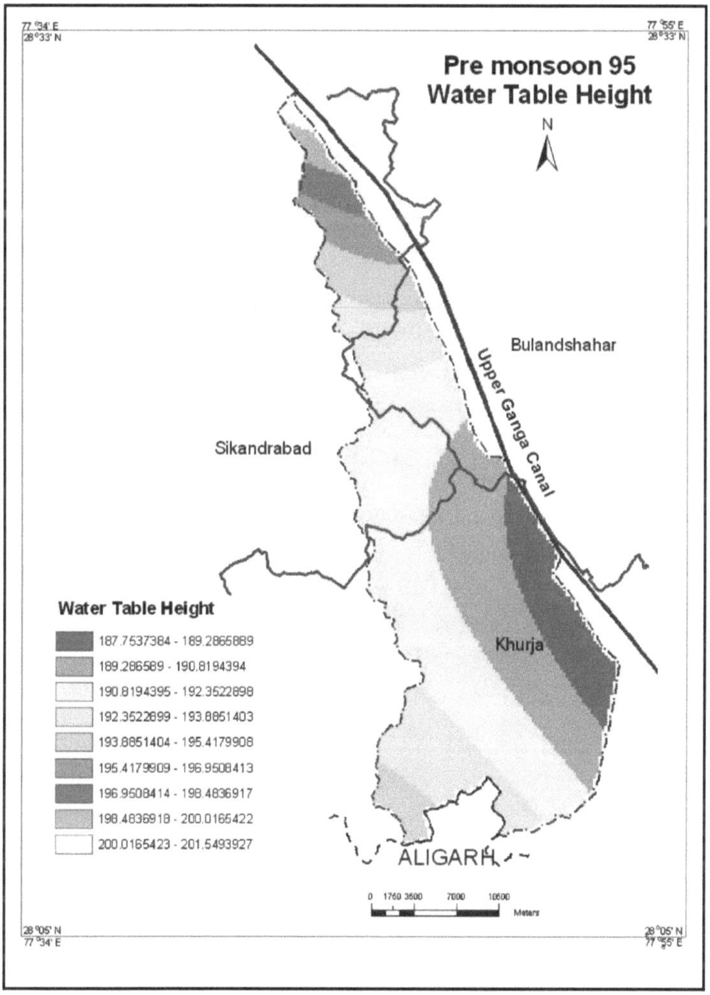

Figure 14. DEM of pre monsoon water table depth for 1995

Conclusions

In the present paper, application of GIS in groundwater modelling has been discussed with the examples. The applications discussed in this paper demonstrate that GIS based groundwater modelling can save time and labour. GIS can be used for map based groundwater modeling using Darcy's law at the defined flow boundaries and continuity equation to the enclosed polygon. GIS can also be linked with groundwater models through an interface program. Interface programme like *Con2grid* can be used to interface the groundwater model, prepare the input files and process the output of the model, prepare the perform the various operations like conversion of polygon to grid, assigning the values of various model input parameters to each grid, linking the model with GIS and displaying the model output. Successful application of ArcInfo interface *Con2grid* to prepare the input file for PMWIN to predict the water table fluctuation in the command of Dadupur distributory in Bulandshahar district of Uttar Pradesh demonstrates its potential in groundwater modeling.

References

Boniol , D., Munch, D. and Williams, M. 1993. Mapping recharge to the floridsan aquifer using geographic information system. In : Proc. 13th annual ESRI user conf. (Vol. 3), Palm Springs, CA. 24-28. May, 1993. Environmental Systems Res. Inst. Redlands, C.A.

Crown, D.L. and Wagenet, R.J. 1996. Application of GIS to the modeling of non-point source pollutants in the vedose zone: A conference overview. *Journal of Environmental Quality*. 25: 403-411.

Fayer, M.J., Gee, G.W., Rockhold, M.L., Freshley, M..D. and Walters, T.B. 1996.

Estimating recharge rates for a groundwater model using a GIS.

Goodchild, M.F., Parks, B.O. and Steyart, L.T. 1993 (eds) In: Environmental modeling with GIS. Oxford University Press.

Harries, J., Gupta, S., Woodside, G. and Zeimba, N. 1993. Integrated use of a GIS and a 3-dimensional finite element model: San Gabriel basin groundwater flow analysis flow analysis In : Goodchild, M.F., Parks. B.O. and Steyaert, L.T. (Eds.). Environmental modeling with GIS. pp 168-172., Oxford University Press, New York.

Johnson, S. Gary, Cosgrove, M. and Joe Spinazolla. 1998. Use of MODFLOW for development of response functions. MODFLOW 98, GOLDEN, CO., International Groundwater Modeling Center and Colorado School of mines. www.if.uidaho.edu/SR3/MODFLOW98.html>05/12/2002.

Krabbenhoft, D.P., and Babiarz, C.L. 1992. The role of groundwater transport in aquatic mercury cycling. *Water Resources Research* 28 (12) : 3119-3128.

Maidment, D.R. 1993. GIS and hydrologic modeling . In : Environmental modeling with GIS (eds.) Goodchild, M.F. , Parks, B.O. and Steyert, L.T. Oxford University Press, New York. 147-167.Maidment, David R and Kris Martinez (1996) Groundwater flow modeling using Arc View. Centre for Research in Water Resources, University of Texas at Austin.

http://www.ce.utexas.edu/prof/maidment/gishydro/africa/ex5af.htm. http://www.ce.utexas.edu/prof/maidment/gishydro/docs/reports/ye/sect4.pdf

Mane, M. S. (2004) GIS based modeling of surface and groundwater supplies for conjunctive water use planning. PhD Thesis, Division of Agricultural engineering, Post Graduate School, IARI, New Delhi

Mane, M.S., D. K. Singh, A. K. Singh and A. K. Bhattacharya (2004) GIS interface for groundwater model MODFLOW. In: *Proceedings of the Symposium on Geoinformatics Applications for Sustainable Development*, IARI, New Delhi, and pp 103-109.

McDonald, M.G., and A.W. Harbaugh. 1988. A modular Three-Dimensional Finite-Difference Groundwater Flow Model-Techniques of Water Resources Investigations of the United States Geological Survey., Book. 6, Chapter, A1, U.S. Geological Survey.

Pinder, G.F. and Gray, W.G. 1977. Finite element simulation in surface and sub-surface hydrology. Academic Press, New York, pp 295.

Prickett, T.A. and Lonnquist, C.G. 1971. Selected digital computer techniques for groundwater resource evaluation. Bull No.55, Illinois State Water Surv. Urbana, pp 62.

Prickett, T.A. 1976. Advances in groundwater flow modeling. In :Saleem, Z.A., Advances in Groundwater Hydrology. American water well Association , Minnestova, U.S.A., pp 102-112.

Rogowski, A.S. 1996. GIS modeling of recharge on a watershed. *Journal Environmental Quality*. 25:463-474.

Rushton, K.R. and S.C. Redshow. 1979. Seepage and groundwater flow. John Wiley and Sons, New York, pp 339.

Rushton, K.R. and Tomlinson, L.M. 1988. Operating policies for a surface/ground water system. *Water Resources Bulletin*, 17 (3): 406-413.

Storetz, M.W., and Bradburry, K.R. 1989. Mapping recharge areas using groundwater flow model - A case study. *Groundwater*, 27(2) : 220-228.

Tsou, Ming-Shu and Whittemore, D.O. 2001. User interface for groundwater modeling: Arc View Extension. *Journal of Hydrologic Engineering*, 6(3) pp. 251-258.

Watkins, D.W., McKinney, D.C., Maidment, D.R. and Min-Der Lin. 1996. Use of geographic information systems in ground water flow modeling. *Journal of Water Resources Planning and Management*. 122 (2): 88-96.

Wilson, J.P., Inskeep, W.P., Wraith, J.M. and Snyder, R.D. 1996. GIS based solute transport modeling applications: Scale effects of soil and climatic data input. *Journal of Environmental Quality* 25 : 445-453.

Ye Zichuan , David R. Maidment and Daene C. McKinney (1996) Map-based surface and subsurface flow simulation models: An object-oriented and GIS approach, CRWR, University of Texas.

http://www.ce.utexas.edu/prof/maidment/gishydro/docs/reports/ye/sect4.pdf

Glossary

Geographic Information System: An organized collection of computer hardware, software, geographic data, and personnel designed to efficiently capture, store, update, manipulate, analyze, and display all forms of geographically referenced information.

Digital Elevation Model (DEM): Representation of the topography of the Earth or another surface in digital format, that is, by coordinates and numerical descriptions of altitude. DEMs are used often in GIS. A DEM may or may not be accompanied by information about the ground' cover, hi contrast withtopographical maps, the information is stored in a raster format. That is, the map will normally divide the area into rectangular pixels and store the elevation of each pixel, hi that sense, digital elevation model (DEM) data are sampled arrays of surface elevations in raster form.

Interface Programme: A computer programme to link GIS with model and process the input and out files. -

Groundwater Modelling: A simplified physical or mathematical representation of aquifer system to study and understand the groundwater behaviour.

Groundwater Simulation: A complete execution of groundwater model including preparation of input file and generation of output file.

❑❑❑

GEOINFORMATICS APPLICATIONS IN AGRICULTURE, 2007
Eds. : A.K. Singh and U.K. Chopra; pp. 293-323
New India Publishing Agency, New Delhi (India)
E-mail : newindiapublishingagency@gmail.com
Web: www.bookfactoryindia.com

11

Participatory GIS for Planning Livelihood Improvement – A Case Study in Canal Command Areas of Bihar

B.Saha, R.P.Jones, A.K.Sikka, P.R.Bhatnagar & S.Chowdhury

Key words : Participatory GIS (P-GIS), Geo-Participatory Rural Appraisal (Geo-PRA) Social Mapping, Resource Mapping, Seasonality Maps, DGPS.

Participatory GIS

Geographical information System (GIS0 and Geographics Information Technologies (GITS) are no more restricted to the elite group of the society. Planners are busy in searching ways in merging community participation with GIS in the context of urban planning, environment equity, sustainable development etc. (Aitken and Michael, 1995, Craig and Ehwood, 1998; Talen ,2000 and Sieber, 2000). In all the cases, common thing is the linking of community participation and GIS in a diversity of social and environmental contexts. Through development, demonstration and modification of methodologies by academics and practitioners the concept of Participatory GIS (P-CIS) emerges. It is an outcome of a spontaneous fusion of participatory forms of development planning with modern information technologies. Experts gave considerable thoughts on the probing questions like whether a bottom up GIS could be successfully developed, how many community participation could be incorporated into GIS and to what context such participation would serve only to legitimize conventional top-down decision making. From these reflections only the concept of people's participation GIS arose.

People of close proximity with common project interest and shared resources, form a community. Through people's participation community is empowered and input of community is critical for defining local issues. People's participation plays major role in community planning and community-developed solutions of a particular issue tend to be reasonable, realistic and sustainable.

A critical aspect of P-GIS projects is the actual form of participation and not the hardware / software configuration. The context of P-GIS is intricately linked to the nature of the participation itself. It is comprised mainly of participatory development research and planning method with GIS-GIT interface (Craig and Ehwood, 1998).

By combining a range of spatially referenced data, information medium and analytical tools, GIS technology enables civil societies to prioritize issues understand them consider alternative and reach viable conclusion.

P-GIS is a system-based process with focus on participation. Although the system will vary greatly from situation to situation, it should be based around identifying user information need and providing this information to support their decision – making. It is designed to provide information to all the diverse stakeholders like farmers, SHGS, scientists, planners and Govt. officials. This is an additional attribute of GIS, it allows the information to be effectively stored, analyzed and prepared for discrimination in a means approximate for each stakeholder group (Jordan, 2002).

Participatory GIS methods

Geo – PRA

Geo-PRA or Participatory GIS is essentially the application of GIS framework and field service techniques to Participatory Rural Appraisal (PRA) exercises (Larroquette and Bhalla, 2003).

Uniqueness of this exercise is that, in this method PRA exercise is conducted on the basis of geo-referenced map where information generated from the exercise is easily stored, updated and temporal change of the resource status can be monitored and compared in an interactive manner.

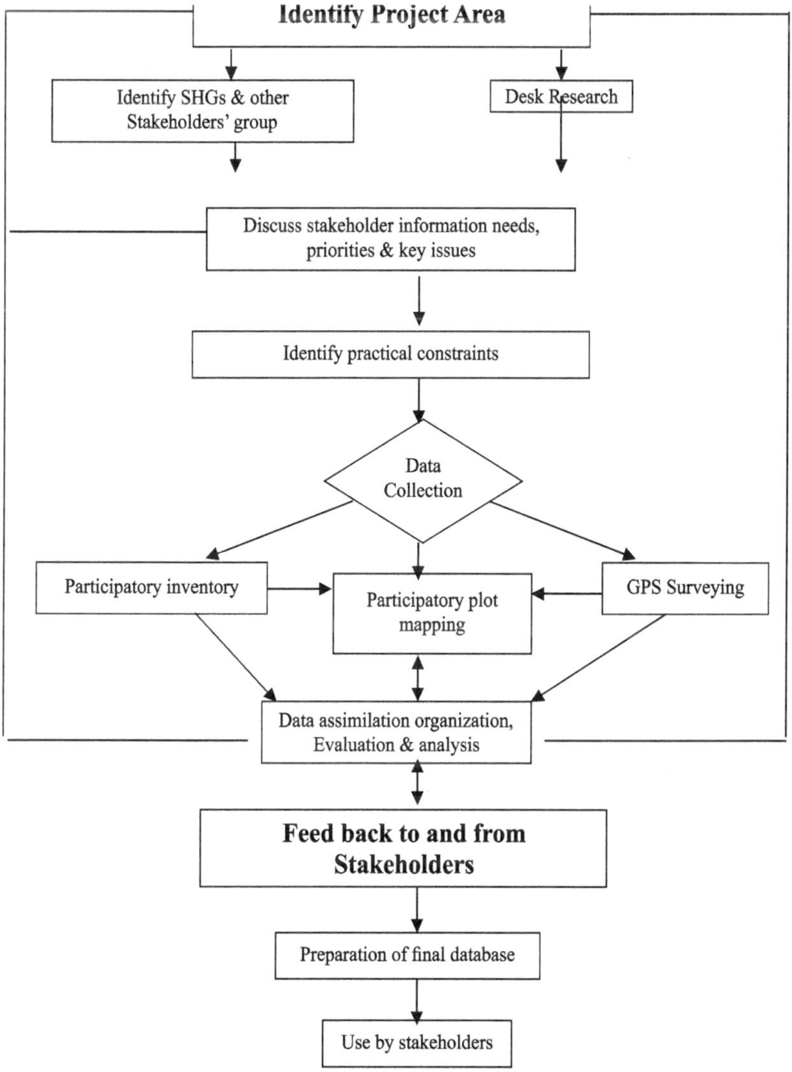

Figure 1. A systematic methodology for sustainable livelihood development and P-GIS

It has following components

- Transect walks and compass traverse
- Social mapping
- Resource mapping
- Seasonality
- Resource use/control
- Micro-planning

Transect walks and compass traverse

This is a participatory survey technique, which is used to develop a walkthrough map of the village. Transect walks typically are taken in (approximate) straight lines with directions being chosen so as to cover a desired resource or region of the village. Many walks can be taken in a village to build a reasonably good overview. With the availability of cheap GPS units, doing so is trivial. However, fairly accurate maps can be generated even with a compass. Important thing is that the data can be uploaded to GIS and used later.

All one needs to do during a compass traverse is to try and travel in a straight line (and direction) until there is a turn. On any one turn a direction (and straight line) should be choosen that covers the longest distance before the next turn. It should be remembered that it is an approximate survey. For improving accuracy in work the reading should be taken by looking at an object in the desired direction and then walking towards that object. Another object should be chosen in the next turn.

Social Mapping

This exercise is normally done to build a baseline of the project villages. The exercise consists of three specific phases.

- A participatory mapping exercise is undertaken during which all the streets are mapped along with various village landmarks.
- This drawing is then transferred to a notebook and the various streets are labeled.

- Finally, a set of key village resource persons accompanies the survey team to each street where a checklist of details about each household is filled.

Linking social mapping to a GIS

There are two basic ways of attaching spatial attributes to households (the normal unit of a social map).

- Each house should be assigned a unique ID and it should be located either using a GPS or by mapping (compass traverse).
- The household values should be assigned that corresponds to other areas in the village, which have already been mapped. For example plot number to assign them land use, resource locations for resource mapping etc.

Resource Mapping

This exercise provides a good baseline of the natural resource use and pattern of a village or a project area. It is a map of the boundaries of the village and enables the study of land, water, cropping patterns and other resources of the village.

The process

- A cadastral or revenue map of the village needs to be scanned and digitized.
- Landmarks of the village, such as well, roads, hand pumps etc. are marked on the map. These points may be collected either via a GPS unit or during transect walks or other surveys.
- The digitized map is printed on a large paper sheet. The larger the better, but also the more expensive.
- A PRA like exercise is then organized where the people of the village are requested to draw the map of their village on the floor based on the digitized map brought along.
- They then carry on drawing a classic Resource Map of their village on the floor.

- Once the map is completed, it is transferred onto the digitized paper map.
- The map is then entered in the GIS software for archival and analysis.

Seasonality

This exercise is useful to assess the various time patterns of the village activity. It helps to study seasonal variations for example rainfall, credit, availability of labour, agricultural operations and recurrent sicknesses.

The process

- This is typically done at the beginning of a traditional PRA exercise. Usually the resource persons chosen for this exercise should be knowledgeable about the village.
- The twelve months of the year are represented with stones or any other material.
- Issues are then identified and described on paper sheet used for the exercise or on the ground.
- A stick proportionate to the amount of rainfall may represent rainfall for each month. This can yield surprisingly accurate data.
- The activities of the village can also be represented through symbol. For example sowing season, peaks of agricultural works, seasonal sicknesses and unemployment times appear clearly.

Use of seasonality maps

This exercise is done to capture critical information pertaining to the pattern of activities taking in the village along the year. When are the men busy and when are they free to do other jobs? What are the good times for asking for contribution either in kind or in cash according to the income and availability of labourer? The rainfall pattern can be very useful when other means of collecting rainfall data is not available.

Resource use/control

Rresource use and control exercise provide useful insights into the manner in which communities partition available resources. This may be done through different exercises like ballot boxes.

Using GIS it is possible to convert a resource use and control exercise from a perception based process to a process that is based on factual spatial data. To do so, the facilitator is required to have completed the resource mapping and prepared the required map. This provides the respondents two kinds of information.

- Prioritizing of the resource is possible on the basis of its extent and potential.
- Stakes are more clearly articulated and any negotiations are based on factual data and not wrong perceptions.

Typically, a resource use/control exercise is preceded by the group having a discussion on a map or, in the absence of one, the creation of a resource map. The group is then able to take up specific resources for the use/control exercise and subsequent discussions are more focused.

Micro-Planning

In this process developed resource map is used for micro level planning which normally refers to the level of detail where every nitty gritty is factored into the plan.

Micro-planning exercises are normally conducted with primary stakeholders. However, in large projects, which can involve hundreds or thousands of primary stakeholders; the level of facilitation is often limited to representatives of these stakeholders. Further, in situations where there are multiple stakeholders from different socio-economic backgrounds, a micro-plan can involve a period of negotiation and ratification by a larger village body such as the Panchayat or Gram Sabha.

Participatory GIS technologies

Application of GPS/DGPS

Global Positioning System (GPS) is a combination of hardware and software having three main segments: space segment, control segment and user segment. Space segment consists of satellites, including back up satellite, orbiting earth at high altitude (approx. 20,200 km). Twenty-one satellites are placed in near circular or orbit at a height of 26,600 kms semi - major axis.and each satellite completes one rotation in twelve hours of side real time. Control segment is the main part of GPS system that monitors the space segment i.e. the satellite's transmission of navigation messages and sends signals for adjustments, if required, and controls the satellite systems continuously. Besides that it determines the GPS systems time and the behaviour of satellite clocks. Third component is the user segment, which is being utilised for different applications. Real time GPS is known as Differential Global Positioning System (DGPS). It plays an important role in collecting timely, reliable, unambiguous geo-referenced spatial data of large area in a cost effective manner. Main component of back held DGPS are antenna/receiver, solar panel, and battery unit with charger, system unit and handheld computer. Receiver unit receives the signal from the satellite and sends it to the system unit, which in turn, passes it onto the handheld through communication port. Data is recorded in the hand held computer, which is downloaded to laptop/desktop computer for further processing

Conceptual models for management of multi-resource inventories namely water structure, forestry, vegetation, wild life, rare and invasive species, cultural and geographical features, outdoor recreation, conservation areas etc. by employing GPS along with other information technology tool, samples design, spatial statistics under client-server architecture were developed by workers at different places. Notable application areas in forestry are: area estimation for timber, inventories, archaeological sites, diseases and pests sites, identification of genetically superior trees, delineation of forest stands, navigation to sample points, road/trials, fire perimeter mapping and wildlife movement monitoring (Evans *et al.*,1992; Rodgers *et al.*, 1996). The biggest obstacle in the use of GPS signals in forestry is canopy interference and multipath errors. Studies have been conducted under variable forestry canopies, terrain and satellite availability evaluate the GPS accuracy (Deckert and Bolstad, 1996). Result indicates a post-processed accuracy ranging from 3 to 6 m, the later under worst condition. Studies at the Austin Cary Memorial Forest, University of Florida, have resulted in sub-meter accuracy when navigating to sample points. This can be attributed to the relatively flat terrain, close proximity to Coast Guard real time differential correction signals, canopy structure and good mission planning. Yule *et al.*, (1999) developed a acquisition system to monitor in-fields performance of an agricultural tractor in combination with a real time DGPS link. Palmer-Jones (2000) studied farm size, land tenure and productivity in Bangladesh using DGPS along with GIS technology. Saha *et al.*, (2001,2004 and 2005) showed the potential use of DGPS in cadastral mapping in canal command of Bihar and eastern U.P.

Case Study of canal command of Bihar

Background

Agriculture is the backbone of rural economy in the eastern India. The region comes under high potential low productive irrigated ecosystem. Though the region is bestowed with highly fertile soils and abundant water resources, it is considered as high potential low productive zone. Inefficient and inequitable water management characterizes canal irrigation in Patna canal system. On farm water management finds the problem at distributary level typified by top end bottom end problem, poor development of field channels, lack of land leveling and lack of co-operation between farmers. Agricultural and irrigation research and development follows a top-down technocratic approach focusing on improved agricultural technologies suitable for elite farmers rather than institutional constraints. Water Users Association (WUA) ignores water management problems like unreliable main system supplies &

unidirectional externalities from upstream distributaries in the main system. Non-irrigation water use like aquaculture has been ignored. Consequently agricultural productivity is low leading to high level of poverty. Agricultural productivity enhancement through integrated land and water resource management is imperative for livelihood improvement of rural areas. In a DFID/NRSP funded project for participatory development of sustainable, pro-poor, gender and environment aware land and water use technologies and institutions; cadastral mapping was undertaken to develop plot GIS for initiating dialogues between different stakeholders for planning livelihood improvement by addressing the above mentioned issues in a comprehensive manner in RP channel 5 distributary command area of the eastern Indo-Gangetic plains.

Materials and methods

The Study Area

The study area comes under agro-ecological region 13. Fieldwork was conducted at different villages of Right Parallel Channel 5 (RP Channel 5) command area (latitude 84^0 51′ - 84^0 55′ E and longitude $25^0$25′-$25^0$28′N) of Patna Canal System. RP Channel 5 is the 7 th distributary of 6 km length and carries a discharge of 1.42 Cumecs. Another sub-distributary Tangrela takes off from 5.10 km of RP Channel 5 and of 4.4 in length. Its design discharge is 0.62 cumecs. The combined culturable command area of both is 1400 ha. Sixteen villages namely Amwa, Alipur, Bawan, Bedauli, Danara, Fatehpur, Gopalpur Gangachak, Harpura, Mohammadpur, Nishapura, Rampur, Sangrampur, Sikaria, Telpa, and Unchauri are selected for the study.

Methodology

Three main steps were involved: the first step was to produce geo-referenced database, the second was to use that to systematically and periodically enter data on variables through participatory approach which could be observed and the third step was to initiate dialogues between SHGs, farmers and scientists regarding the future soil, crop and water management strategies on the basis of thematic maps. The interaction envisaged between GIS/direct observation data and other activities are shown in the Fig.1.

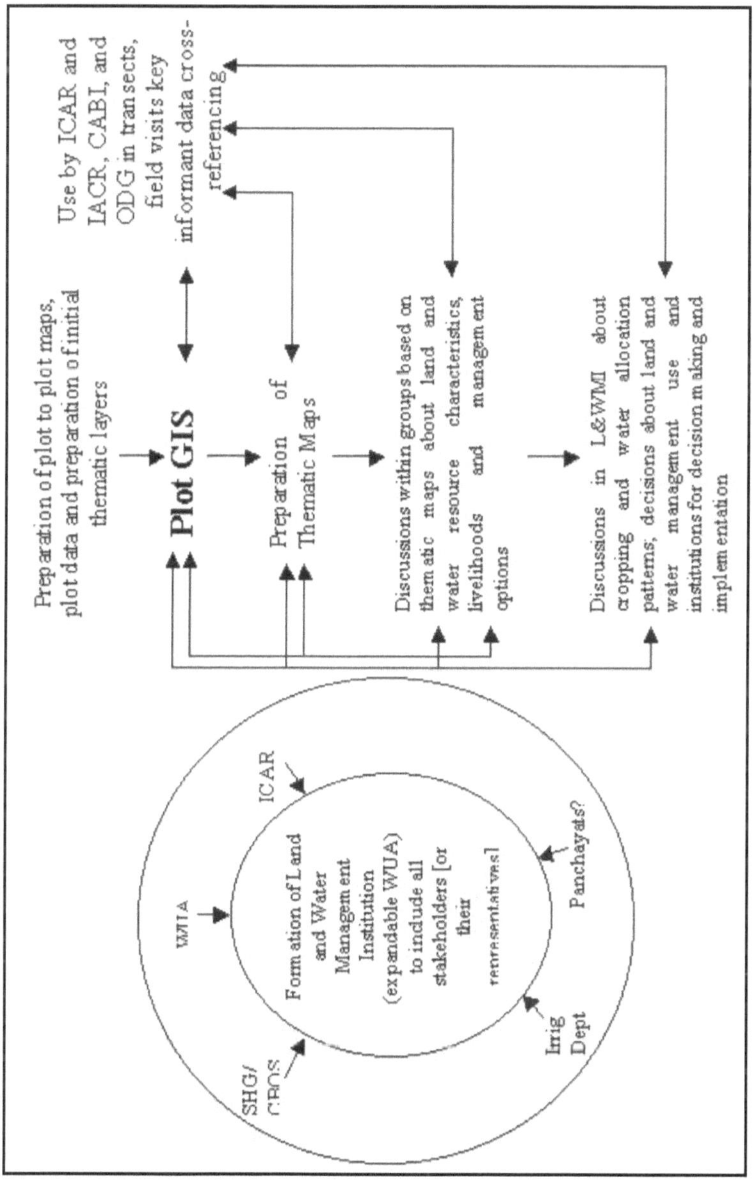

Figure 1. Interaction between GIS/direct observation data and other activities.

Revenue maps of all the villages prepared during colonial period were collected from the Patna Cadastral survey offices. In the initial participatory sessions with the self-help groups, specific problems were discussed. Various issues starting from canal and water management, waterlogged areas, time of transplanting of crops, tillage options and agricultural laborers were the focus of the discussion. Villagers communicated well with the existing cadastral maps, but the condition of maps was very poor. After a number of discussions, it emerged that; agricultural production could be increased through improved management of land and water, appropriate technological interventions and improved communication between different stakeholders. The information requirements were a combination of basic spatial information and management information for managing land and water resources in best possible ways using qualitative and quantitative information. Data collection process was developed after finalization of information needs.

The boundaries of several villages were walked with the local key informants using a RACAL made Landstar DGPS georefernced maps. A number of prominent features (the Patna Canal Bank and the road from Akhtiarpur to Nabatpur) were traced using the GPS technique (Photo 1). These larger scale

Photo1. Survey of village boundaries

traces were used to reference the 1:50000 survey of India toposheet no. 72 C/15. They were also used to reference a LISS 3 image from the IRS 1D satellite taken in February 2000. Given confusion about village boundaries geo-referencing has been facilitated by an up-to-date satellite image.

Participatory plot mapping

As the plot recorded in the village maps appeared to have poor correspondence with plots identifiable in the field; the plot to plot mapping was undertaken using Differential Global Positioning System (DGPS) and LASER range finder with built in compass coupled to Huskey fex21/Pocket PC (Compaq IPAQ) using pocket GIS software in WINDOWS CE environment (Photo 2). Solar energy was used for charging the instruments in the field through solar panel. Each plot has been given a particular identification number. Direct observation data of the plot regarding land type; land use, soil type, crop, tillage, drainage and irrigation are recorded through a structured schedule formed using Satellite Form software on a Palm IIIxe organiser. Maps and direct observation data are integrated through Arc View GIS software through the common identification number. In that process 22000 plots in the entire area was mapped and cadastral maps with identification number of plots have been constructed.

Pocket PC, Huskey, Laser range finder and Palm used for plot mapping and direct observations recording

Photo 2. Instruments for plot mapping

Photo 3. Mapping with key informants and DGPS

Photo 4. Using Solar Panel for charging instruments in the field

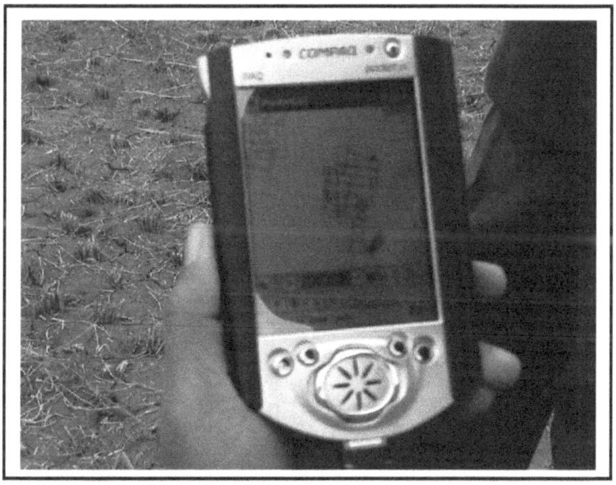

Photo 5. Plotted map on Pocket PC (PDA)

Community Building Organization (Self Help Group) volunteers were trained in using palm IIIxe for recording direct observation data in respect of each agricultural plot as per cadastral maps. Information regarding date of sowing/ transplanting of crops, tillage practices, source of irrigation and methods of irrigation, location of outlets, outlet commands, tube wells and presence / association of different self-help groups in the villages were recorded by them. They charged Rs 0.5 per plot for recording data during temporal and spatial monitoring. They collected response from villagers regarding the maps. Thematic maps of the whole command area regarding different aspects have been produced on the basis of direct observation data and the raw inventory data were made available to use by water management groups, agronomists, social scientists, farmers and self help groups for taking management decision with suitable interventions. On the basis of temporal and spatial monitoring of plots in *kharif* season; date of transplanting of paddy in different outlet commands, area coverage under *kharif* crops mainly paddy in each command were estimated from attribute data of thematic maps. Tube well-irrigated area has been marked and the plots irrigated through tube well during different periods of time have been identified and possibilities of conjunctive water use in different reaches of the canal are being explored. Outlets along the distributary and command areas under each outlet have been marked with the key informants. Using these Geo-referenced plot maps of villages, Participatory Rural Appraisal exercise (Geo-PRA) was conducted to integrate social and natural resource information in respect of each plot. The information were uploaded to the laptop and integrated in a GIS environment to develop thematic maps.

Results and Discussions

Cadstral maps with basic information of land type, soil, *kharif* irrigation source and drainage congestion were prepared (Map below).

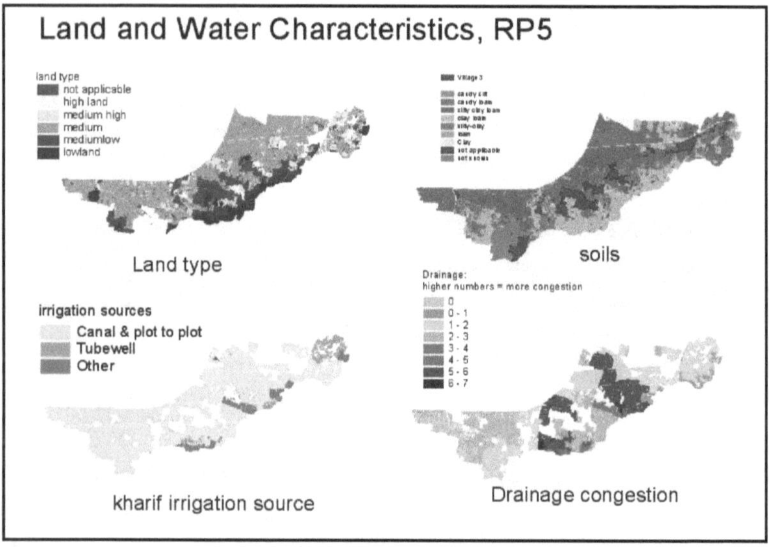

Map

Information extracted from the thematic maps revealed that, the soil texture of the area is dominated by silty clay loam (49%) followed by clay loam (36%) as per the (Map and Fig. 2) Heavy soils were observed in the proximity of drains and low-lying areas.

Initially, the villagers were excited with rapid survey techniques. They were capable to interpret the maps. Gradually, they learnt to use palm computer. Village volunteers themselves later recorded direct observation data using palm and cadastral maps during temporal monitoring of changes in natural resources in exchange of money. In these processes unemployed youth of villages acted as service providers and got employment opportunity.

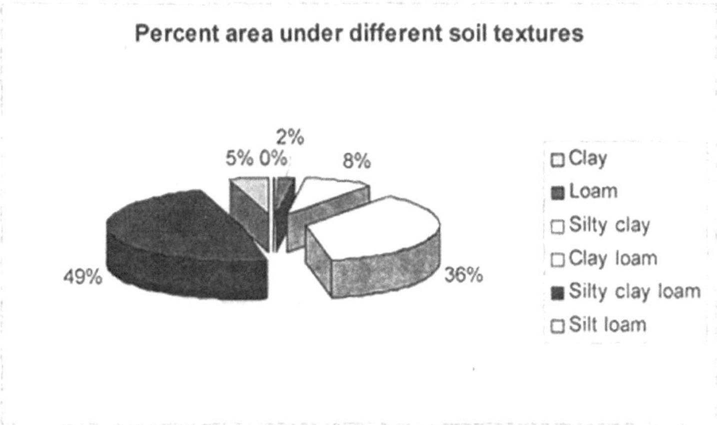

Figure 2

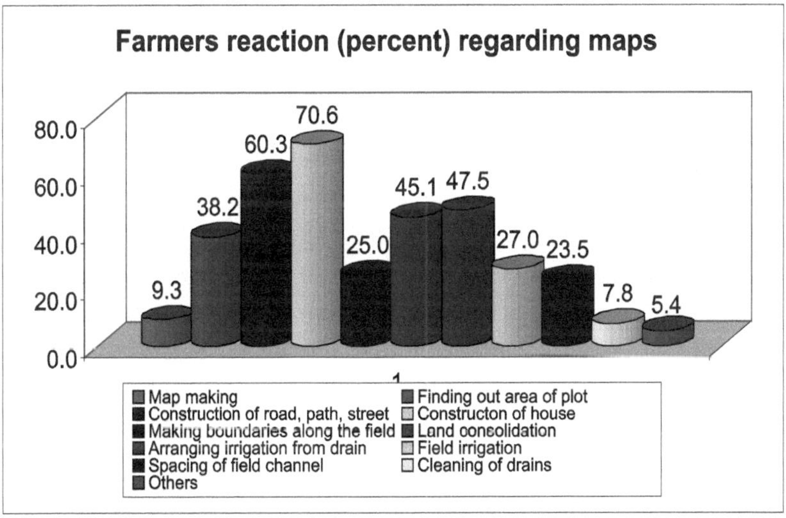

Figure 3

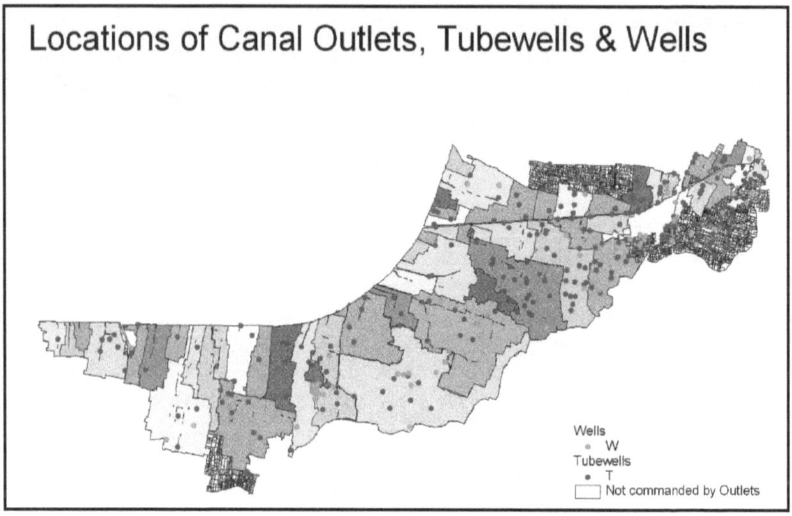

Map

DGPS was much harder for community to use for scientific mapping. However, villagers quickly grasped ideas of DGPS-Laser range finder survey of the plots. Farmers' responses regarding the maps revealed that, the maps were very useful for solving land dispute problems, planning agricultural activities, and judging accessibility of their fields to irrigation (Fig.3).

Total number of outlets existed in above the study area is 55 covering 1276 ha command area (Map). The percentage of paddy coverage varied between 80-100% in most of the outlet *kharif* season 1117 ha area was covered by paddy. Timely (before 30th July) transplanting was done in 40.6% of the area, whereas it was late in 38.6% and very late in 20.8% of the area (Map on page 311 and Fig.5). Percentage of area covered by paddy under different outlets and date of transplanting of paddy were used as input of water balance model to generate different scenarios of water availability.

Wheat is the major crop in *rabi* season occupying 46.4% of the total area is raised in proximity to the distributary during *rabi* season. Wheat, oilseeds and pulses (mixed crop) occupied 12.6% of the area. Farmers are taking more pulse crops in the

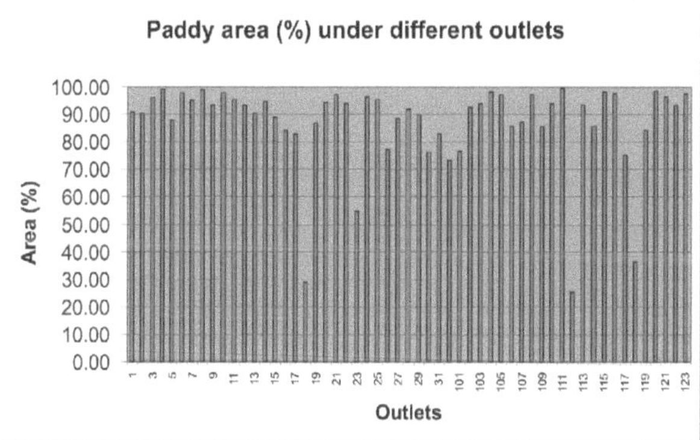

Figure 4

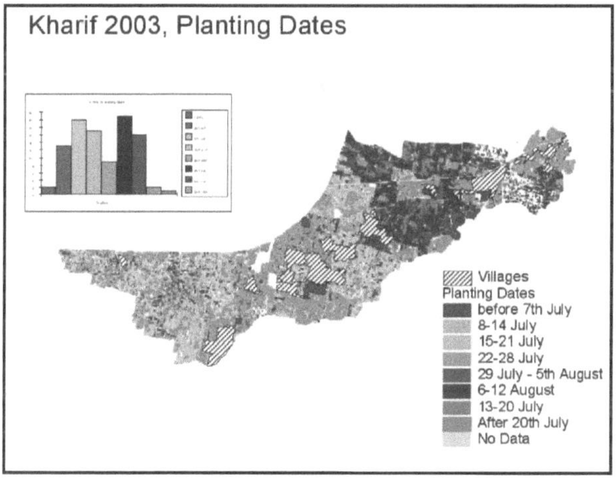

Map

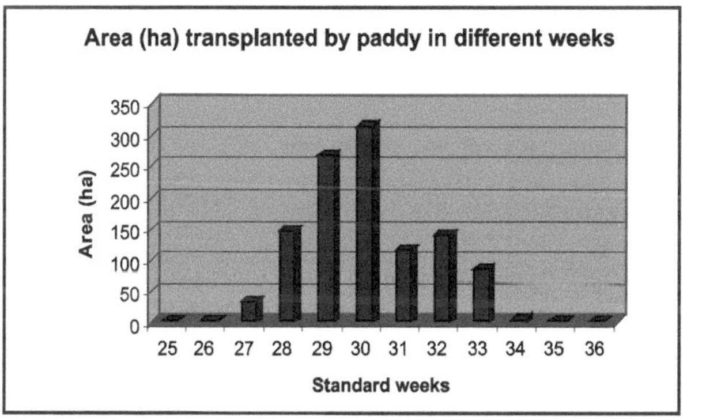

Figure 5

heavy soils of low-lying areas near the drain and lentil occupied 27.4% of total cropped area (Map below).

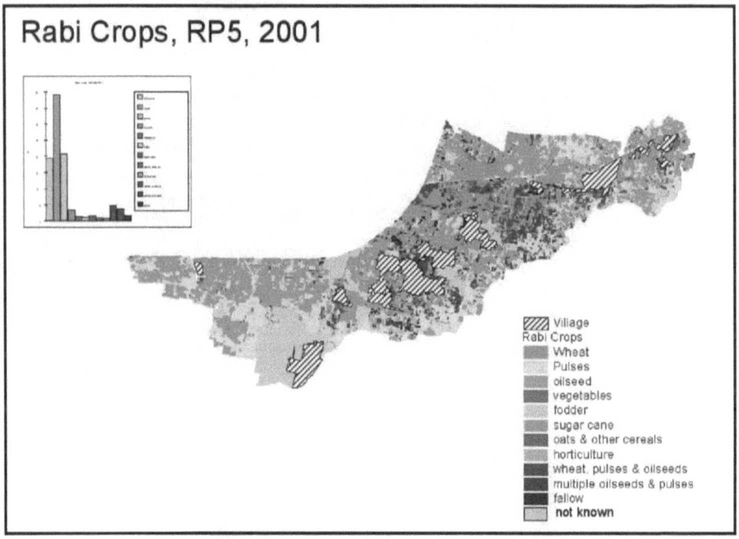

Map

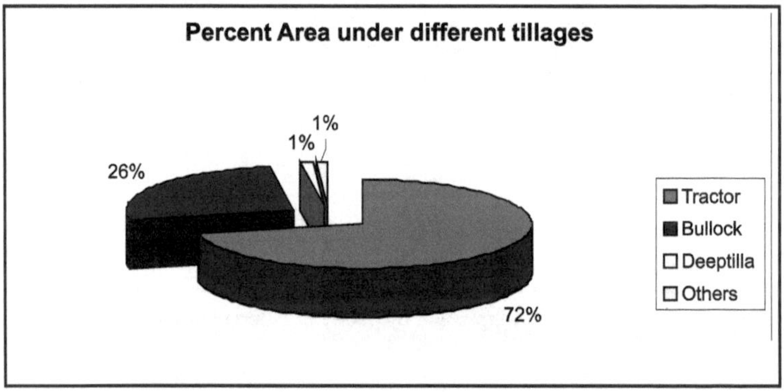

Figure 6

Conventional tillage through tractor is used in 72% of the area followed by bullocks (25%) in the command (Fig.7). Practice of deep tillage and zero tillage is also picking up steadily as per the need of farmers.

In the RPC5 command area, 79% area is irrigated through canal and 8% of area is covered under tube well irrigation. 8%

of the area is covered by both canal and tube well irrigation (Fig.7). 325 wells are existed in the command area, out of which 257 are shallow tube wells, 24 are deep tube wells and 44 are open wells (Map on page 310). Highest tube well density of 46% was observed in tail reach followed by middle and head reach (Fig.8). Overall tubewell density in the command is 25%.

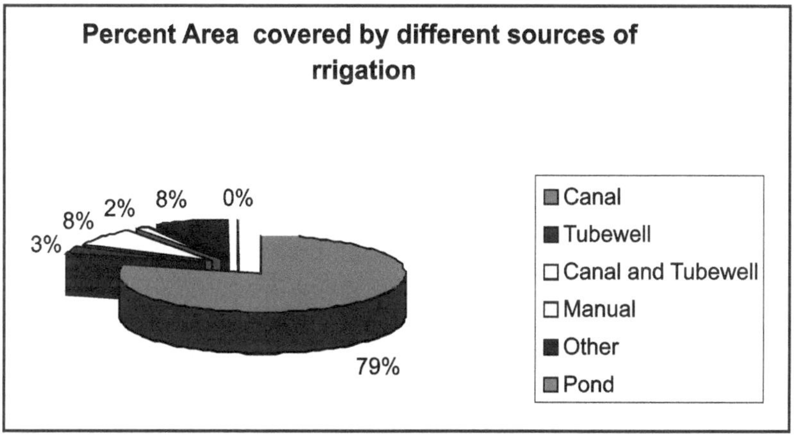

Figure 7

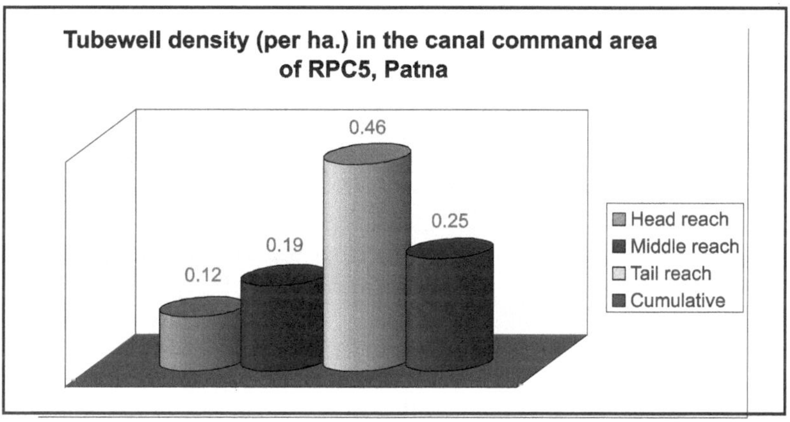

Figure 8

Among different irrigation methods plot to plot irrigation has been noted to dominate with 89% of the area followed by 8% (Fig. 9) through field channels causing loss of nutrients from top soil and poor yield of crops.

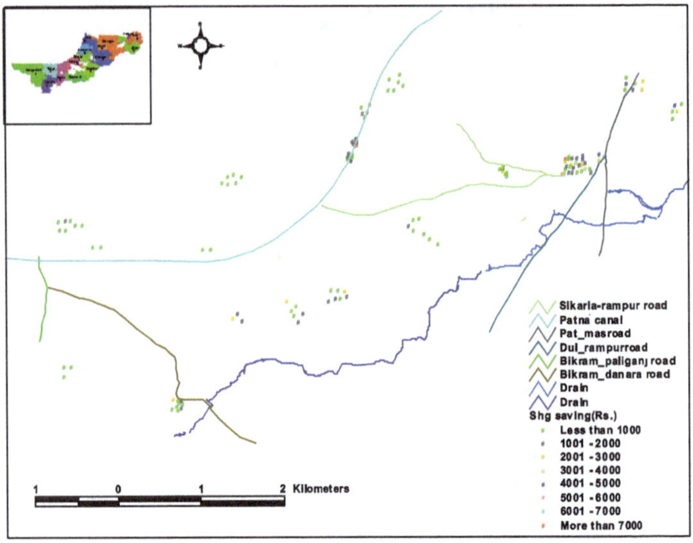

Map Distribution of self help groups and comparison of savings in different villages

Distribution of 150 self-help groups, which were formed on the basis of microcredit and saving systems, were depicted on the thematic map (Map above). Savings of different self-help groups are compared with the thematic map.

Use of Maps

Cadastral Maps in Communication and Planning

Cadastral maps were used for communication among different stakeholders for planning improved management related interventions as shown below:

Use of maps by SHGs

- Maps were shown and discussed within group members for crop planning on the basis of land type, access to irrigation to their plots, soil texture, crops and varietal performances
- Maps were used to substitute conventional rice varieties with hybrid rice towards improved productivity

Scientists discuss land and water management with farmer using rough mosaic of Revenue Maps

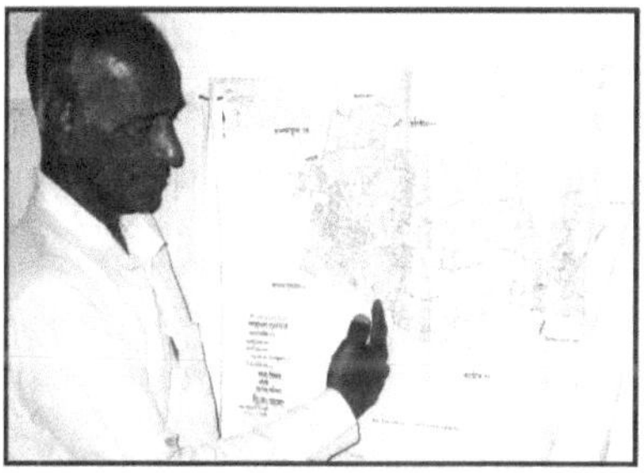

Farmer points to revenue maps to describe land characteristics

Community is using GIS maps in participatory mapping and livelihood planning

- It would facilitate landless people to choose lands in lease judging on the basis of natural resources availability
- These thematic maps contain comprehensive information in respect of agricultural plots including boundary updates, therefore are better than revenue maps which contain area of the land only
- Concentration of self help groups in different villages of canal command are easily depicted and different income generation activities of the group could temporally and spatially monitored and compared
- During the process of developing maps more meetings and discussions with the members of groups are required for growing awareness among the groups and making it user friendly
- The maps have good potential market among the group members

- The maps could be effectively used for improvements of livelihood of civil societies (*Source:* Sunil Chowdhury, CPSL, through personal communication)

This leads to innovation in localised efforts to manage waterlogging, cleaning of well, purchase / rent of tube wells to irrigate vegetables.

Livelihood Analysis

Livelihood analysis was done after one year of group formation. It was observed that

- There are 4000 households in the project area
- There are 32 Social groups
- 50 percent of the households are landless
- 829 persons have salary as an income source
- 522 households are doing poultry and 22 of them are in farm basis
- Dairy done by 1029 households
- There are 250 small shops with same number of households
- There are 454 households doing goat rearing, pig rearing. These activities are limited in social Groups
- Households with following proportions do share cropping: Marginal -958(24%) , Small -298(7%) , Medium -162(4%) , Large-43(1%)
- Out of 4000 households 3400(85%) households are involved in agriculture
- Since 36% of the households are landless share cropper they need small technology for better livelihood
- Wage labourer have basket making activities which is done seasonally
- Wage labourer available for 150-200 days and payment done in cash and in kind

Economic Categories in the command

BLAD (Between Life and Death)

- Challenged, long sickness, alcohol abuser, unable to earn, begging, widow, aged without care

Deficit

- Landless, irregular employment, lacking capital, illiterate, alcohol abuser, family size, 'lazy', lacking social support and opportunities Dependant on son

Self-sufficient

- Land holder, use of land , income from other sources

Surplus

- New assets each year, more land holding (but no salary income)

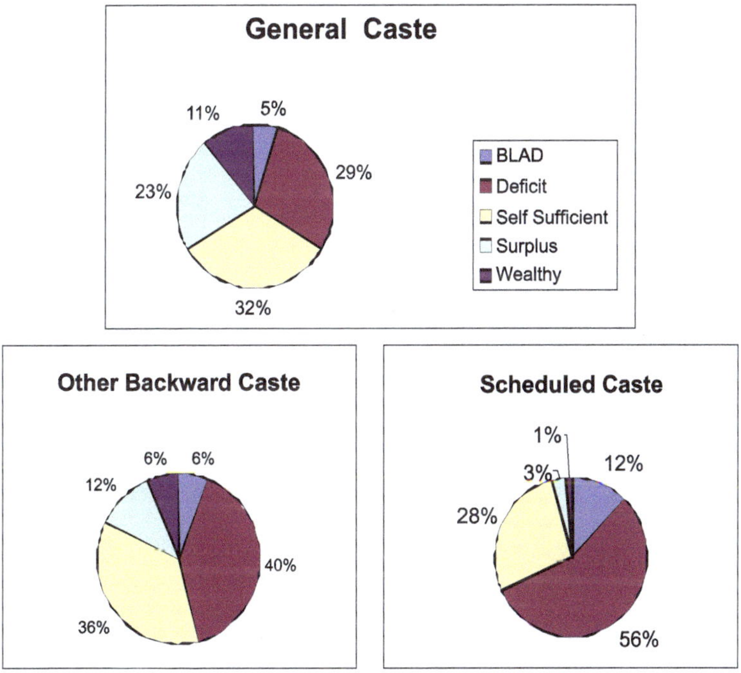

Figure 9. Different economic categories within various castes

Wealthy

- Large land holdings plus income from other sources (esp. salaries)

Assess progress of technical innovations – early transplanting of kharif rice

Land preparation for rice is done only after arrival of monsoon in the region. Delayed rice-planting leads to late harvest and late planting and low yields of wheat in *rabi* season as temperatures climb rapidly in February. Maps helped to monitor progress in adoption of early transplanting with Direct Observation data linked to GIS.

Use of maps for identification of tube well irrigated area for different purposes through direct observations

Tube well irrigated area for different purposes and different periods of time was identified through direct observation data using the cadastral maps as depicted in the following maps.

Tube well water was used for nursery growing in 36 ha of plots. 134 ha of plots were irrigated by tube well water during *kharif* season in conjunction with canal water. Irrigation was provided to 374 ha command area from tube well during water crisis period (Fig.10).

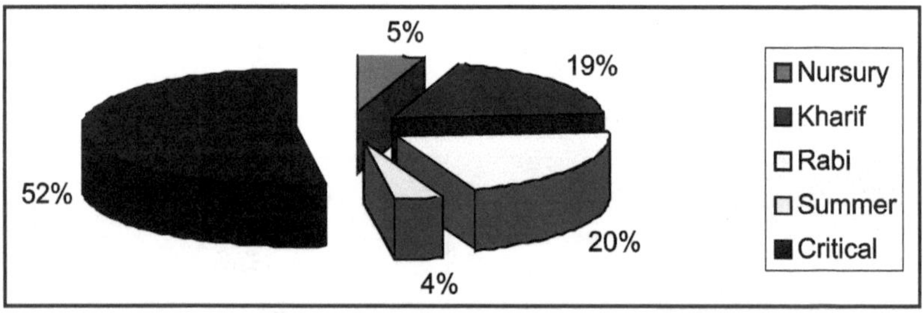

Figure 10. Percent use Tubewell irrigation for different purposes

Addressing Water-management issues

- Water mismanagement through formal and informal outlets has been identified through maps.
- Uncontrolled flow from outlets causes head reach-tail reach problem among community.

- Discussions with Waater Users Association (WUA) with the maps for probable solution lead to the formation of Outlet Mangement Group (OMG).
- Head reach and tail reach problem was mitigated but could not be eliminated.

Community Development

- Maps incorporated into the Dialectic approach for strengthening social and cultural capital.
- Poor and marginalized were empowered and entered into new relationships with scientists and other outside experts.
- WUA/OMG and SHGs recognise mutual benefits from planning water management and cropping patterns, including share cropped lands.

Estimation of water productivity

- Promoted interest in and dialogue on water and land management issues among stakeholders

Thematic maps along with cadastral maps are now being used by water management scientists for estimating water productivity under different outlet commands of RP Channel 5 distributary in terms of plant biomass and food production.

Decision Support System (DSS) tools through linkage of PGIS maps and model

Offline linkage between maps and water balance model has been established to develop a Decision Support System (DSS) for facilitating water management related interventions by outlet management group in cases of different amount of discharges from main canal during on and off periods of canal operation. Thematic maps generated through linkages of Water balance tool and GIS mapping used in Water Balance tool, which estimated that with the uncontrolled outlets, water loss through surplus flow is 32.19 % and water applied to field only 35.38% and predicts that if all the outlets have controlled flow , 25.99% of water loss due to surplus flow and of water available for irrigation rises to 43.52%.

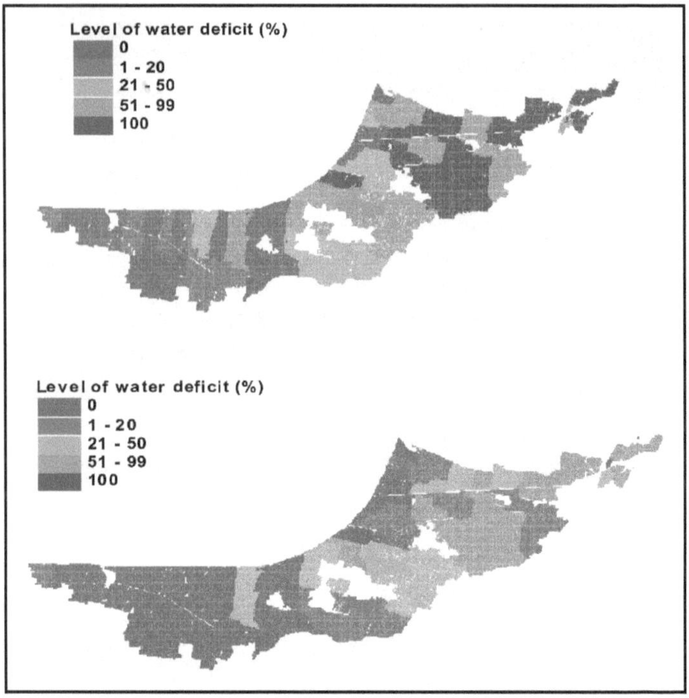

Map

Evaluation of technology

Hardware

- Reliable hardware and software and adequate budget is necessary for the work
- DGPS is not necessary
- Laser Range Finder (LRF), PDA & Pocket GIS, Direct Observation (D.O) & Palm and Satellite Forms are a viable technology
- Regular backup on desktop and data archiving with reliable and properly used UPS is mandatory

Competence in Surveying, GIS and Data Management

- Mapping requires proper aptitude, interest and training

Appropriate management of field survey work during different seasons

- It depends on land characteristics, competence, work organisation and weather
- Visual blockages (trees, sugar cane, buildings, standing crops, standing water, rocky outcrops) creates problem

Participation by community could aid both mapping and Direct Observation

- Community should be organised in advance for proper participation during GIS mapping
- SHG, WUA or Schools/institutes should be trained and used for mapping for cost effective result.

Conclusions

Lessons Learnt

GIS used in participatory processes but not driven by "participatory mapping"

- Most important stakeholders found the GIS useful once it is in place
- But did not appear to recognise potential in advance
- More meetings, interaction and discussions with community are required regarding utilities of maps to ensure proper participation

GIS proved a valuable resource when actors encountered research and communication problems

- GIS needed to be in place to respond to emerging issues and problems

Implications for scaling up above the outlet

- Costs and time to prepare GIS maps was heavier than anticipated. It should be cost effective and rapid, so that interventions should be planned and implemented in time

Scaling Up

- Linkage of GIS and water management model was demonstrated to State level canal managers and promoted dialogue about alternative water management regimens
- Dialogue among project team members and Irrigation department stakeholders using DSS for problems and solution has been initiated for future uptake
- It was presented to Federal Ministry of Water Resources which showed interest in GIS for improved Command Area Development
- Work was presented to different national level and international level symposiums held in India
- Effort was made for scaling up to international agencies through oral presentation in front of water experts at Stockholm water symposium during World Water Week held at Stockholm, Sweden during 21st to 27th August, 2005

References

Aitken, S. and Michel, S. 1995. 'Who' contrives the 'Real' in GIS? Geographic information, planning and critical theory' *Cartography and Geographic Information Systems* 22 (1): 17-29.

Larroquette, B and Bhalla,R.(2003) Field methods in Participatory GIS. Training manual of Founadation for Ecological Research, Advocacy and Learning (FERAL), P.O.Box 28, Pondicherry-605 001

Craig, W. and Ehwood, S. (1998) 'How and why community groups use maps and geographic information.' *Cartography and Geographic Information Systems* 25 (2): 95-104.

Deckert,C.J and Bolstad, P.V. (1996). Global Positioning System (GPS) accuracies in eastern U.S.deciduous and conifer forests. *South. J. Appl.Forestry* 20 (2): 81-84.

Evans, D.L., Carraway, R.W., Simmons, G.T.(1992). Use of Global Positioning System (GPS) for forest plot location. *South J.Appl. Forestry* 16 (2): 67-70.

Jordan, G. (2002) GIS for community forestry user groups in Nepal: putting people before the technology. In "Community participation and Geographical Information Systems" Edited by Craig et al. and published by Taylor and Francis, London and New York. pp. 233-245.

Palmer Jones, R. (2000). Farm size, land tenure and productivity: an approach using Geographical Information Systems in Bangladesh: a report to ESCOR.(Unpublished report.)

Rodgers, A.R., Ramphal, R.S., Abraham, K.F. (1996). A GPS based telemetry system. *Wildlife Soc. Bull.* 24 (3): 559-566.

Saha, B., Das,P.K and Singh, S.R. (2001). Application of Differential Global Positioning System (DGPS) in land water management. Proceedings 16th Convention of Computer Engineers held at Patna during December 15-16, 2001 pp: 131-136.

Saha,B., Sikka,A.K., Jones,R.P., Prasad,S.S, Prasad, L.K. and Singh, R.D. (2004) Application of DGPS, GIS, RDBMS and Geo-PRA techniques for planning livelihood improvement in Bihar and eastern U.P. Proc. of the symposium on geoinformatics application for sustainable development held at New Delhi during 17th to 19th February, 2004 pp.110-114.

Saha,B., Bhatnagar,P.R and Sikka, A.K. (2005). Participatory GIS for planning efficient water resource management towards improved agricultural productivity in canal command areas. Proceeding of abstracts The 15th Stockholm water Symposium held at Stockholm, Sweden during 21st to 27th August, 2005 pp.150

Sieber, R. (2000) 'Conforming to the opposition: the social construction of geographical information systems in social movements', International *Journal of Geographical Information Science* 14 (8): 791-793.

Talen, E. (2000). 'Bottom-up GIS: a new tool for individual and group expression in Participatory Planning', *APA Journal* 66 (3): 279-294.

Yule, I.J., Kohnen, G. and Nowak, M (1999). A tractor performance monitor with DGPS capability. *Computers and Electronics in Agriculture* 23(2): 155-174.

❑❑❑

Colour Version of figures given in Black & White in Text

325-340

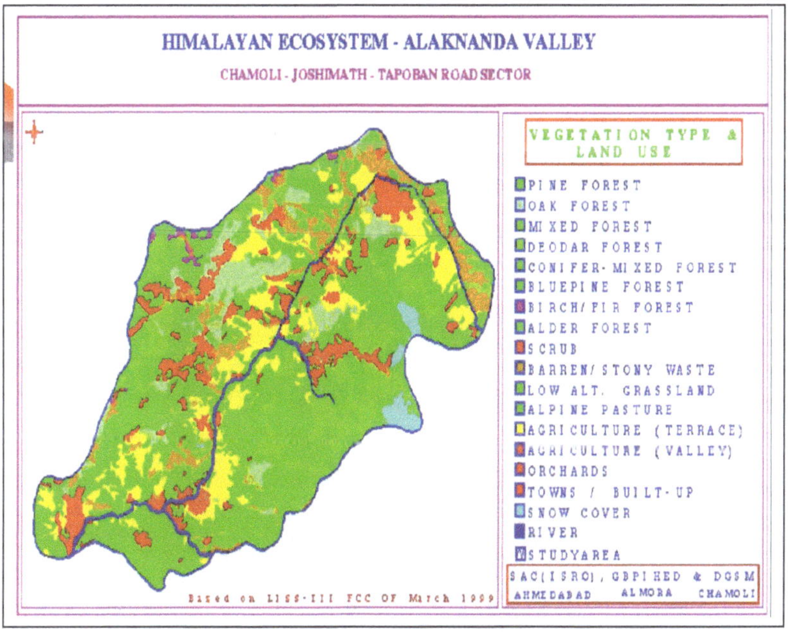

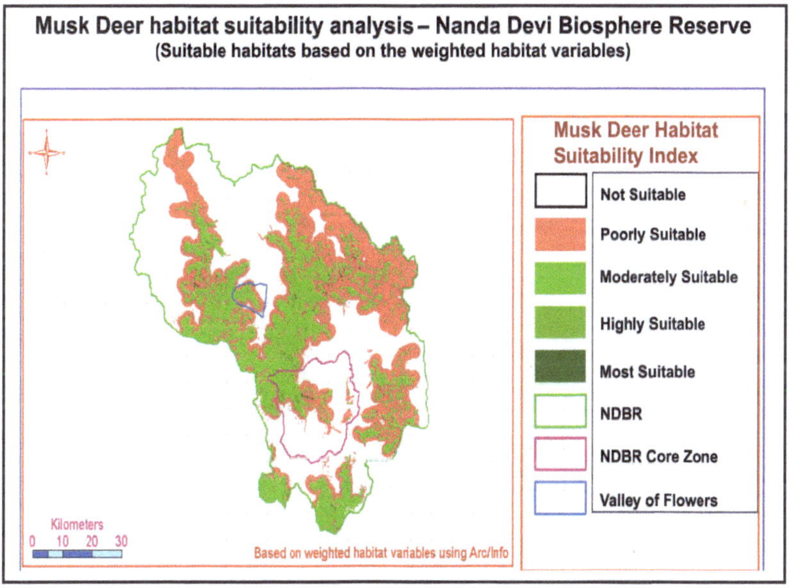

Figures on Page 18

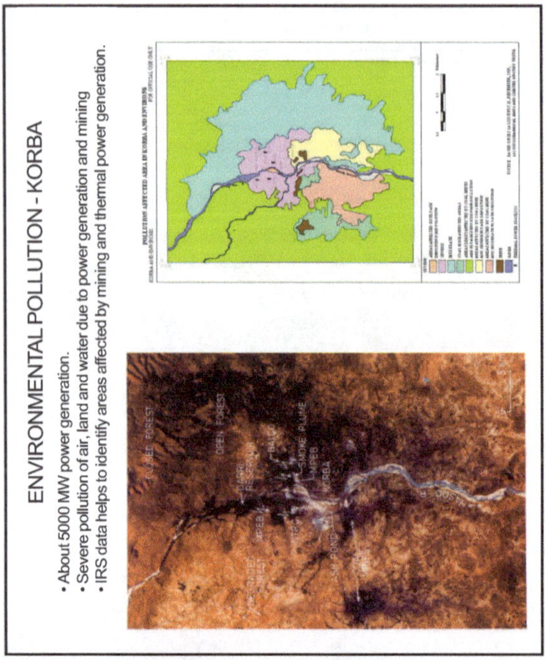

Figures on Page 31

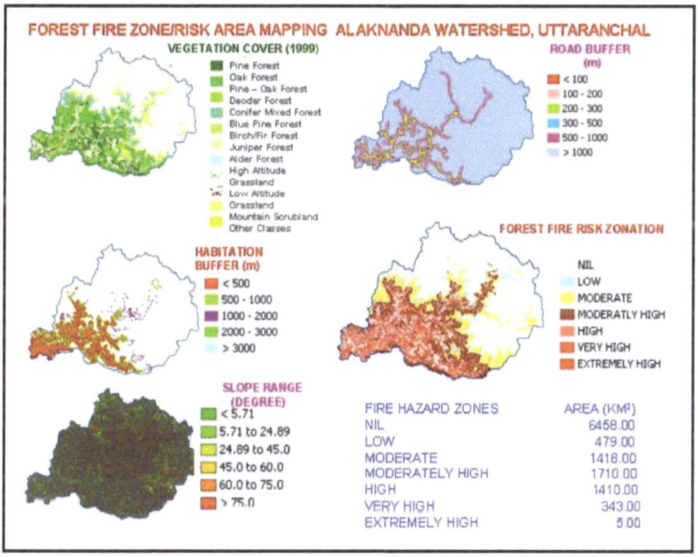

Figure on Page 35

Chapter 2

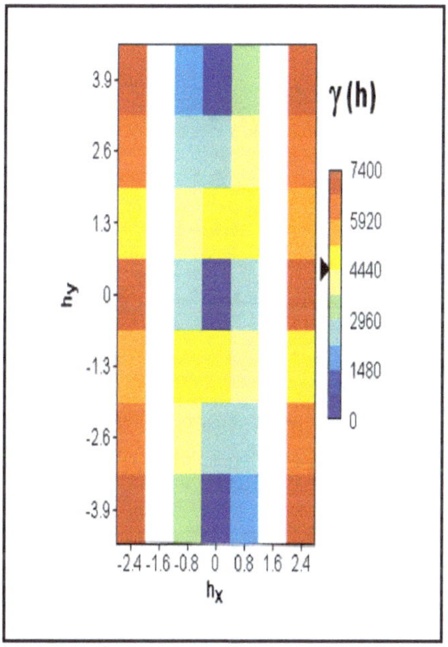

Figure 6. Grid generated for nitrogen availability in the soil
Page 71

Chapter 4

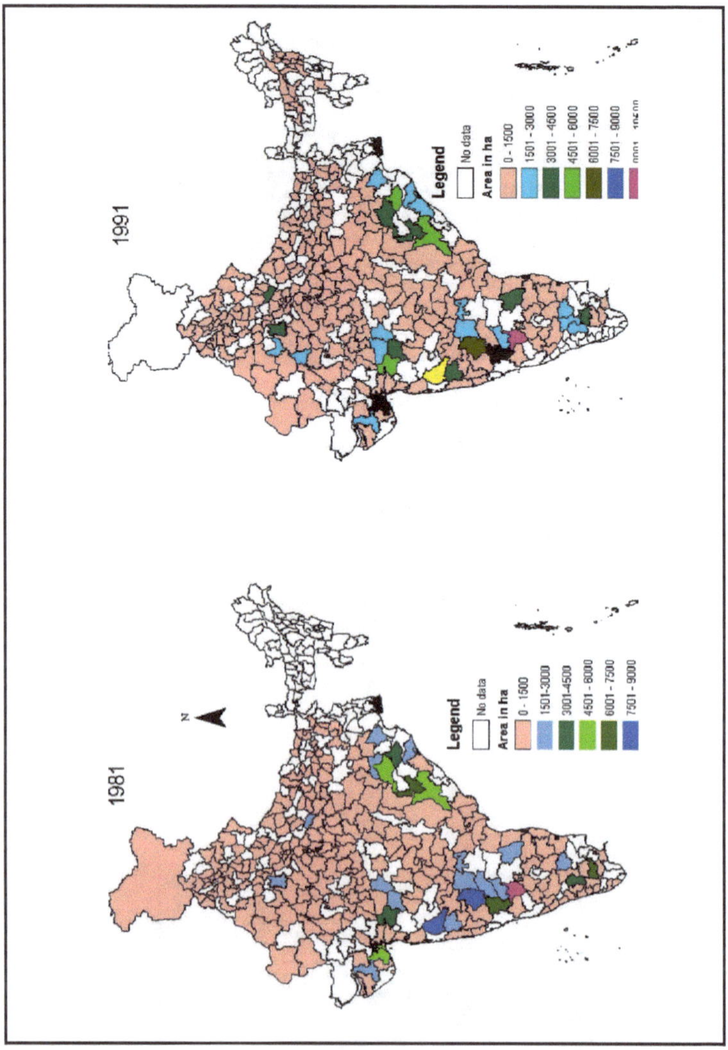

Figure 6. Thematic maps on district wise area of onion in India

Page 116

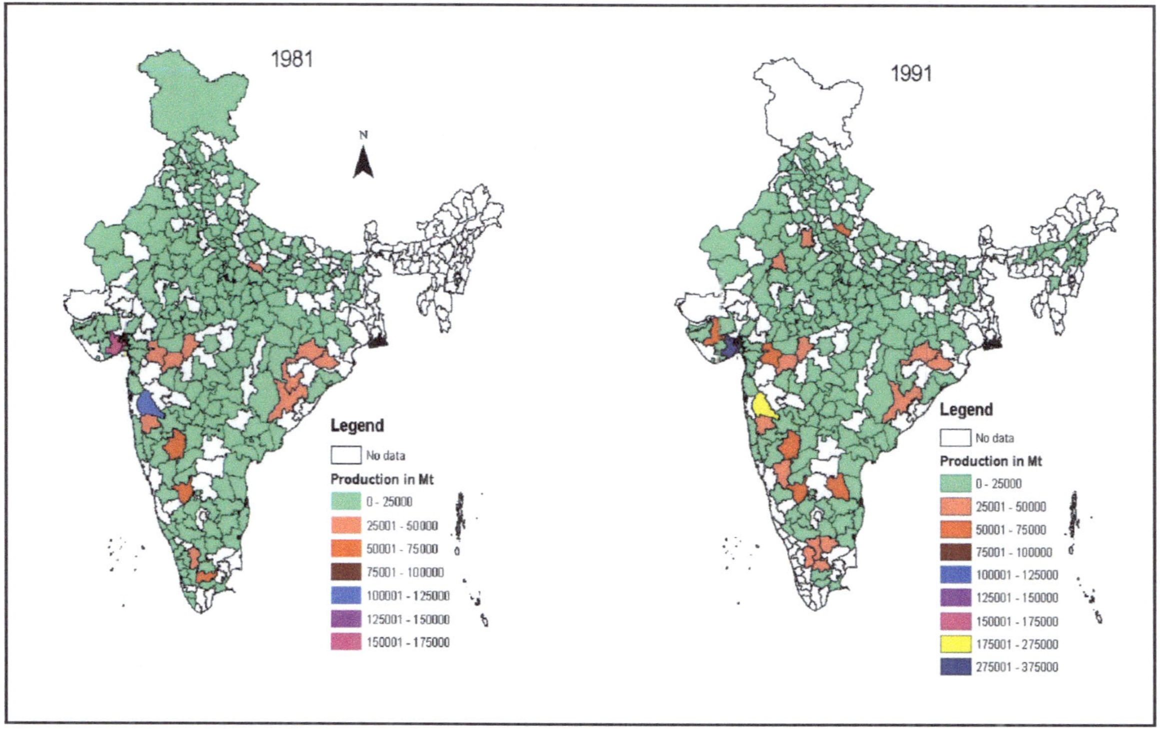

Figure 7. Thematic maps on district wise production of onion in India
Page 118

Chapter 5

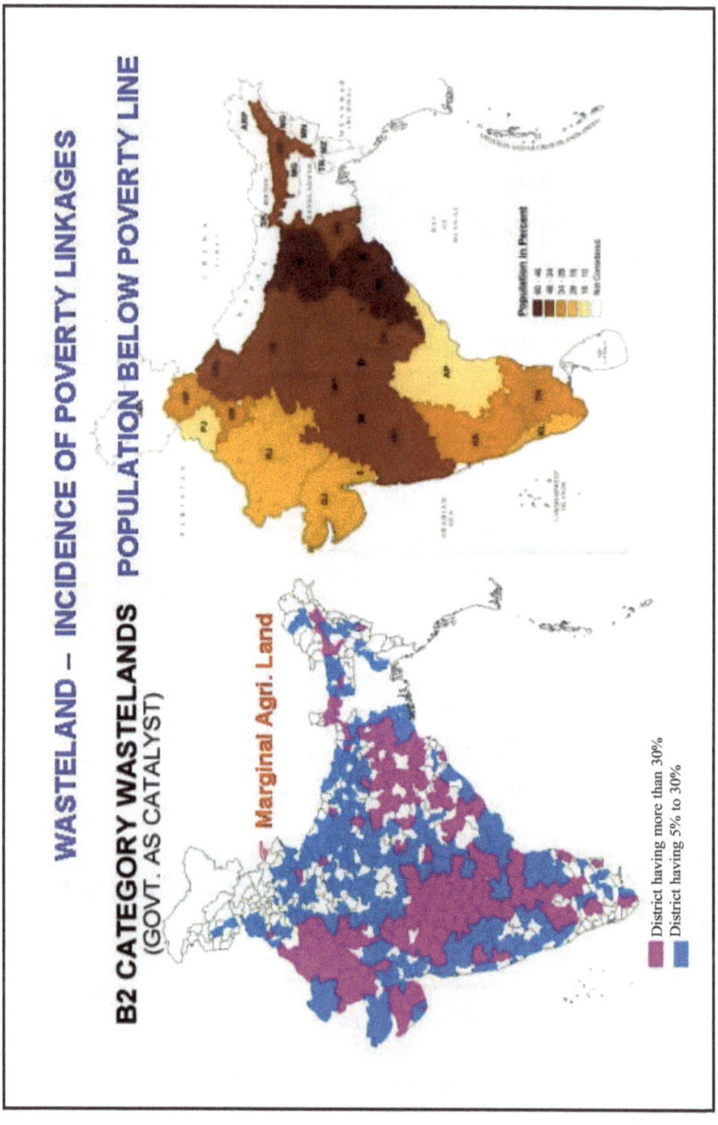

Figure 1. Marginal lands (*Source:* NRSA) and incidence of poverty (*Source:* Food Insecurity Atlas of Rural India) Page 124

Chapter 6

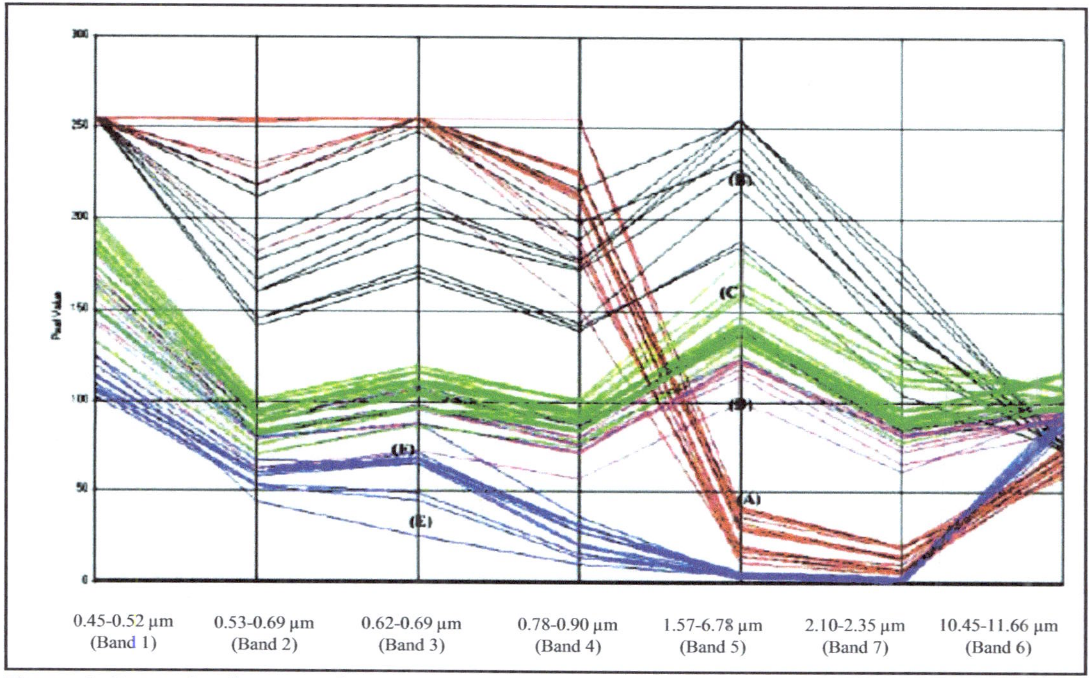

Figure 1. Spectral reflectance characteristics in terms of pixel value based on a September 22, 1992, Landsat TM seven-band data set. Red lines – clean glaciers and fresh snow (A); Black lines – clouds (B); Green lines – recent debris from GLOFs (C); Maroon lines – debris covered glacier (D); Blue lines – clean/melted (E) silty and/or partly frozen water (lake) (F) Page 146

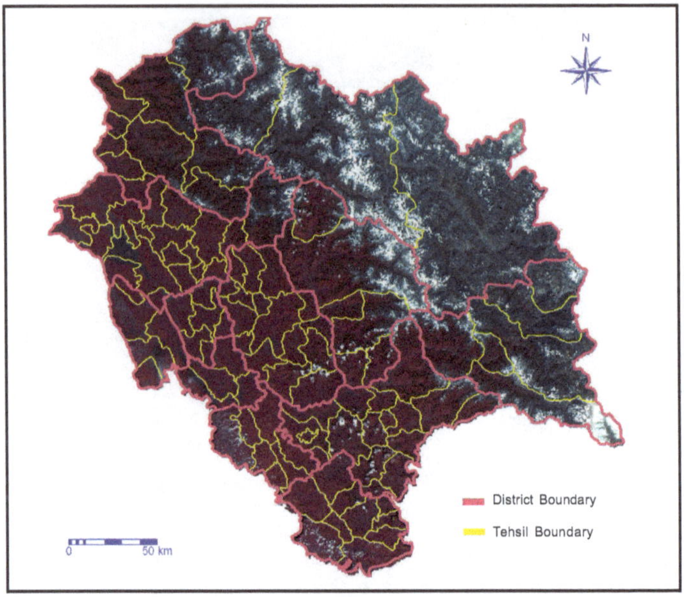

Figure 3. A map of Himachal Pradesh showing Boundaries of Districts and Tehsils. Page 152

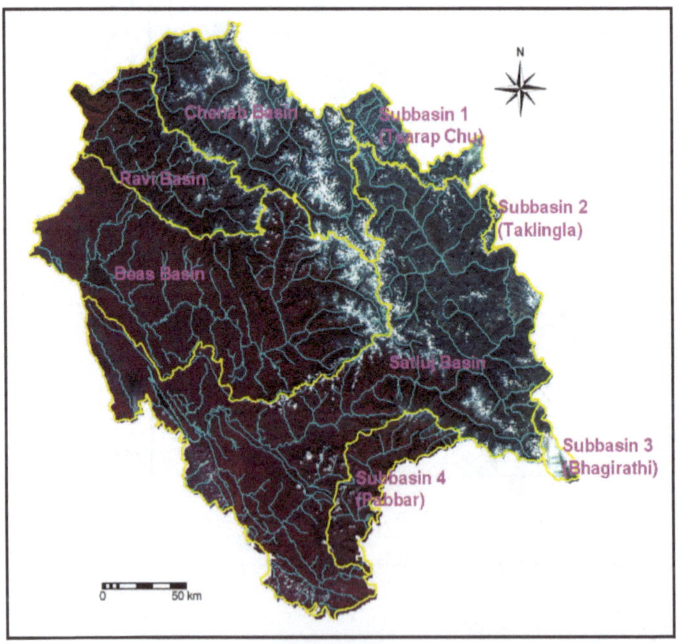

Figure 4. River network and basin wise distribution of glaciers in HP. Page 154

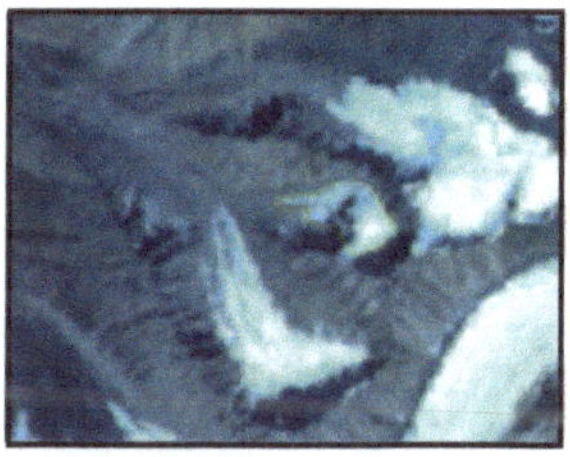

Cirque glacier mapped as Chenab_gr 580

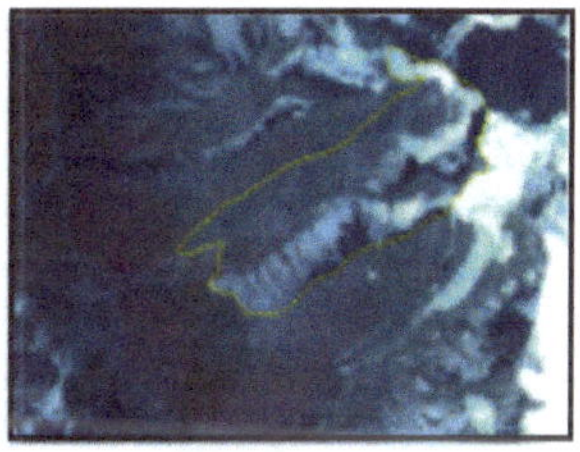

Ice Apron mapped as Chenab_gr 580

Ice cap mapped as Chenab_gr 32

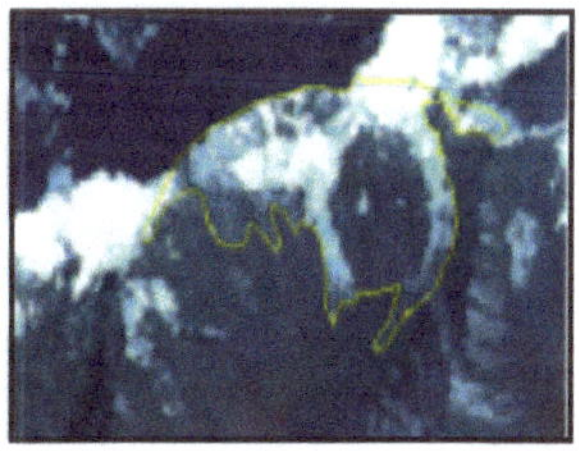

Mountain glacier mapped as Chenab_gr 81

Mountain basin mapped as Chenab_gr 429

Niche Glacier mapped as Chenab_gr 256
identified as Chhangang palu lacier

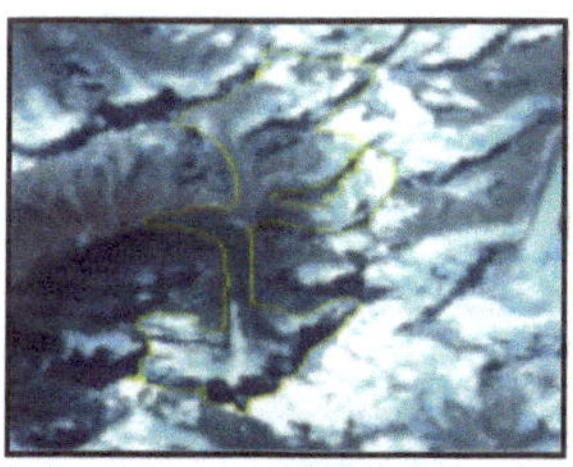

Valley Galcier mapped as Chenab_gr 203
identified as Gangs tong glacier

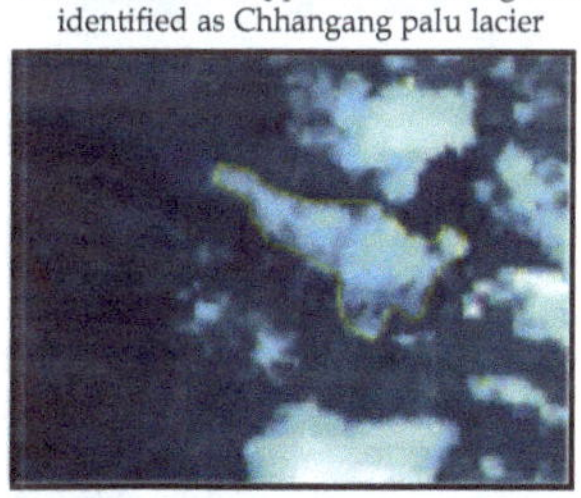

Cirque glacier mapped as Ravi_gr 184

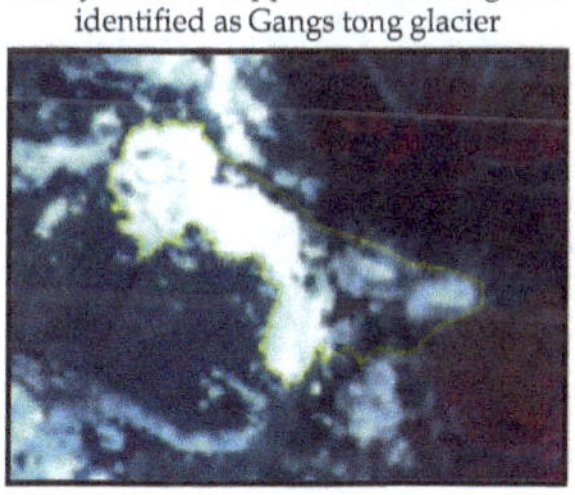

Mountain glacier mapped as Ravi_gr 149
identified as Guwala glacier

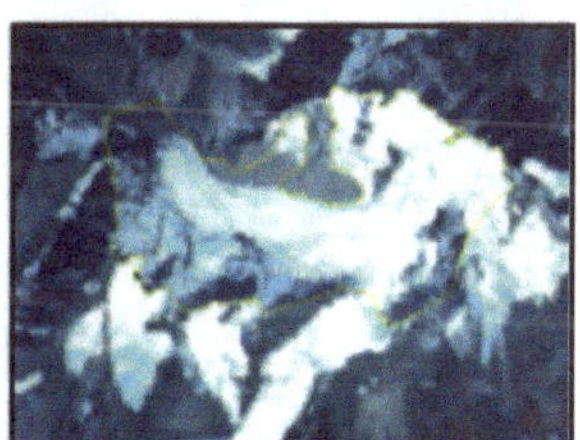

Mountain Basin mapped as Ravi_gr 123
identified as Badunag glacier

Figure 8. Different types of glaciers Page 161

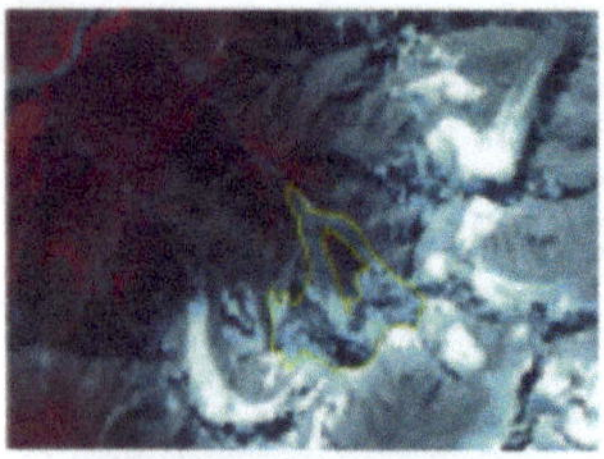

Saro Glacier mapped as satluj_gr 852 - Mountain basin

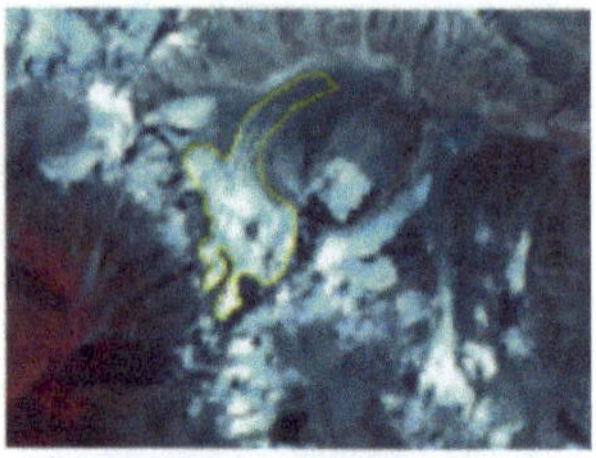

Sankari Glacier mapped as satluj_gr 100 -Mountain Basin

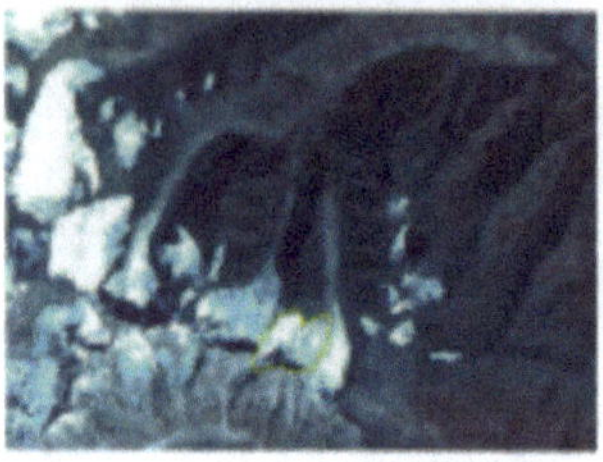

Phuldang Glacier mapped as satluj_gr 109 - Mountain Glacier

Chhota Shigri glacier mapped as Chenab_gr 587 - Mountain Glacier

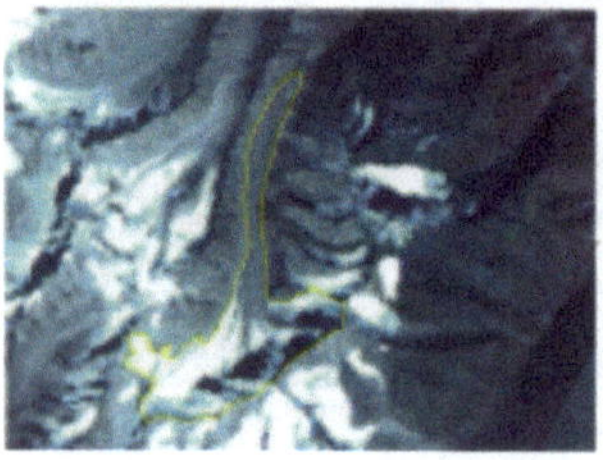

Lambar Glacier mapped as satluj_gr 844 - Valley Glacier

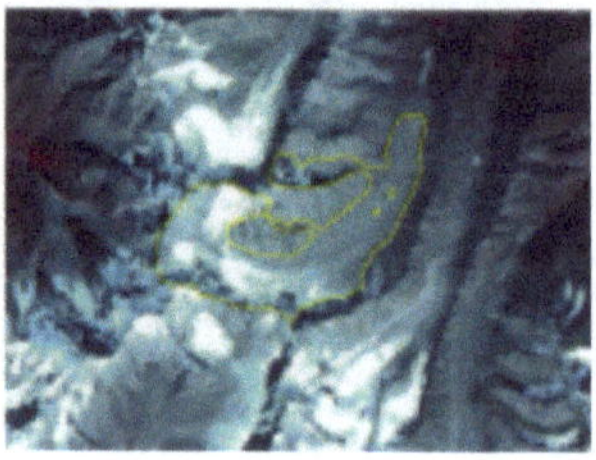

Garo Glacier mapped as satluj_gr 849 - Valley Glacier

Figure 9. Some of the important glaciers mapped and identified using SOI topomaps in Himachal Pradesh glacier inventory. Page 162

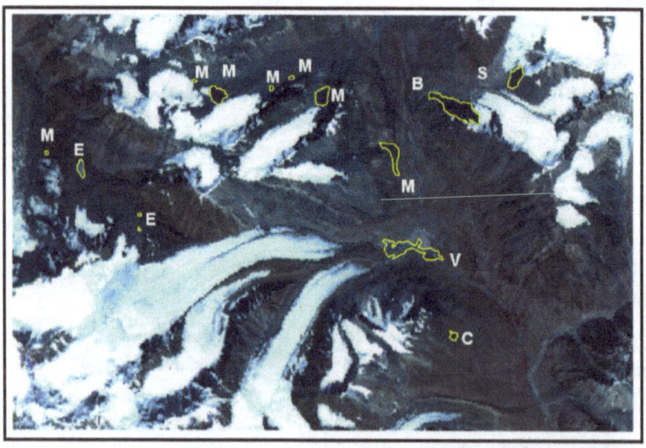

Figure 10. Lakes in the Chenab Basin showing the type of lakes. S-Supraglacial lake, M-Moraine Dammed Lake, E-Erosion Lake, B-Blocked Lake, C-Cirque Lake.

Page 163

Chapter 7

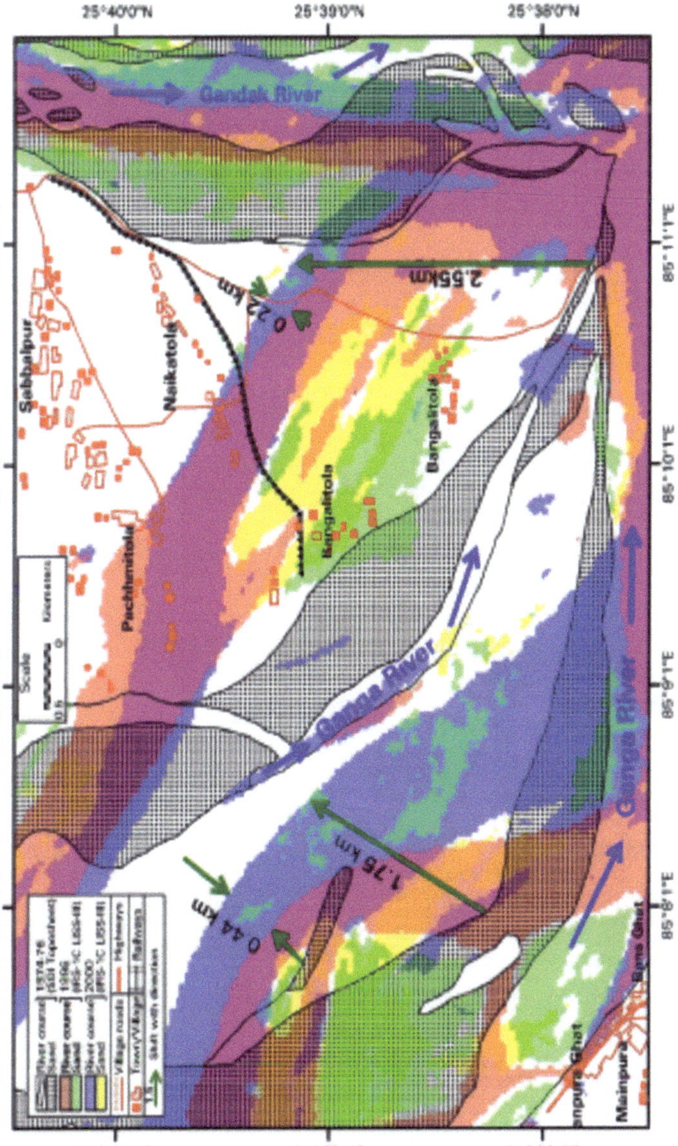

Figure 2: Details of shifting of River Ganga near its confluence with Gandak at Hazipur as obtained from Survey of India toposheets and IRS LISSIII data (1996 and 2000)   Page 187

Figure 3. False Colour Composite (FCC) of Ghatprabha reservoir located in Karnataka (a) October 2000 (b) April 2001 Page 196

Chapter 8

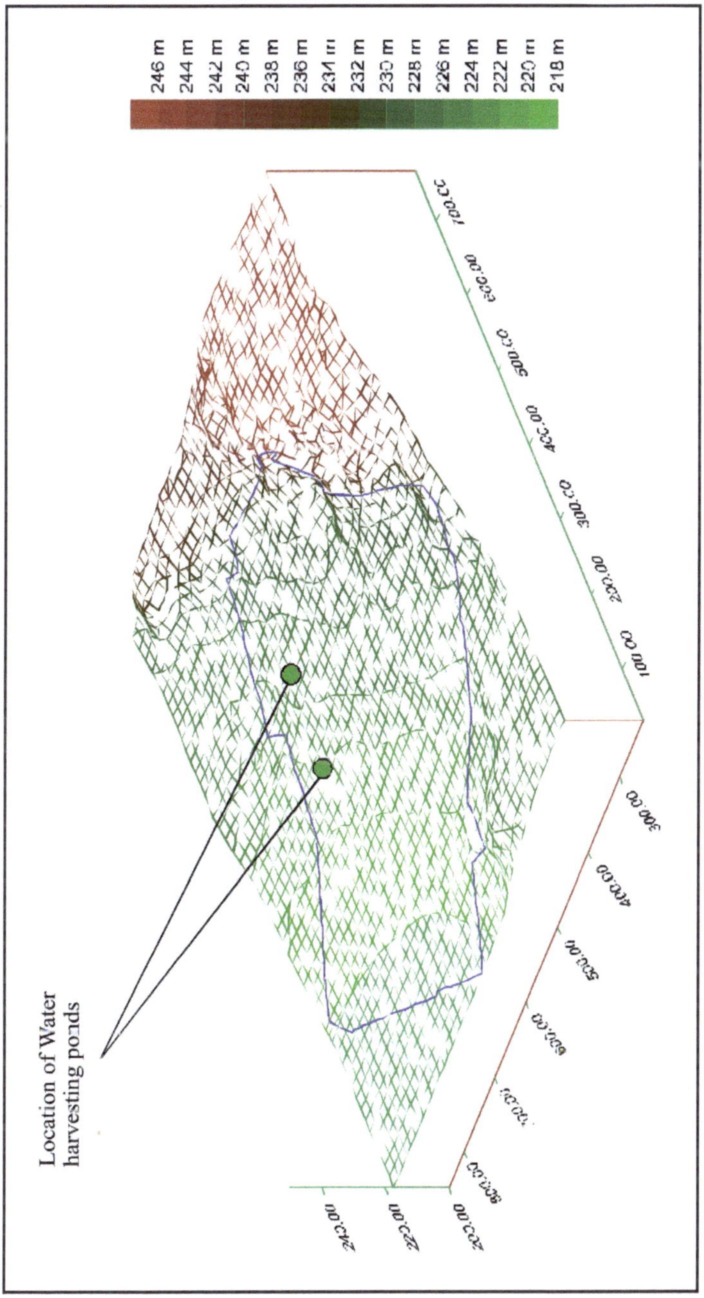

Figure 15. GWIS generated location of potential water harvesting zones in IARI watershed Page 240

Chapter 10

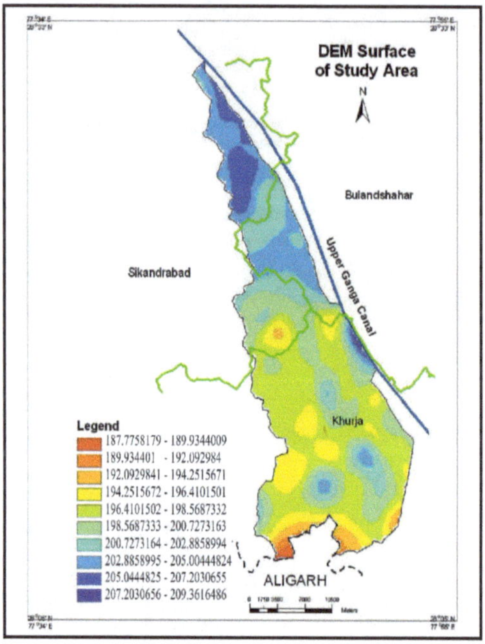

Figure 8. DEM of surface elevation Page 283

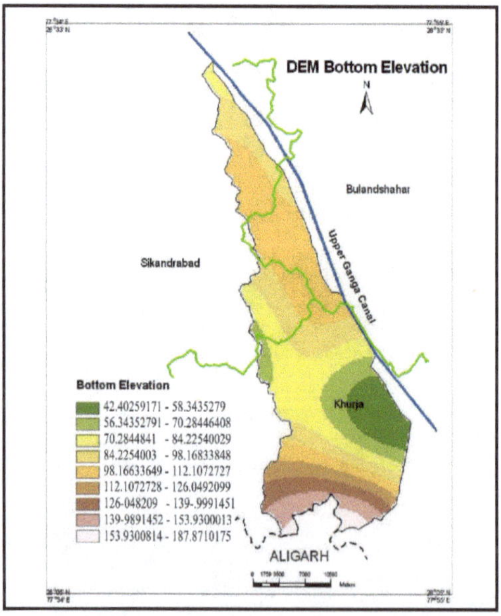

Figure 9. DEM of bottom elevation Page 283

Chapter 11

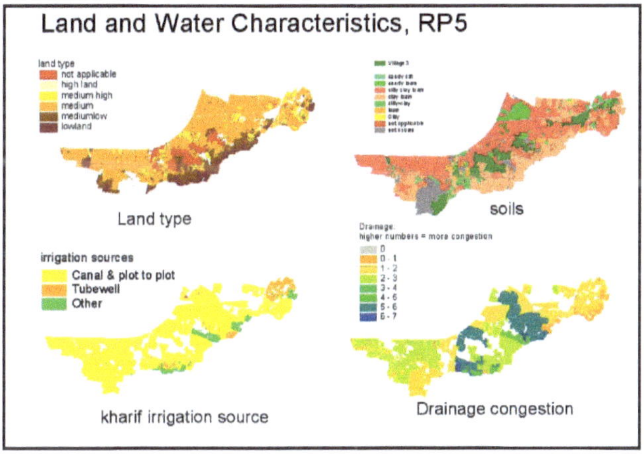

Map Page 308

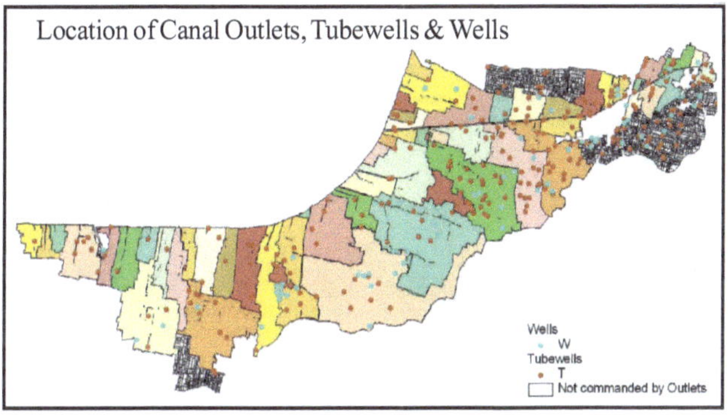

Map Page 310

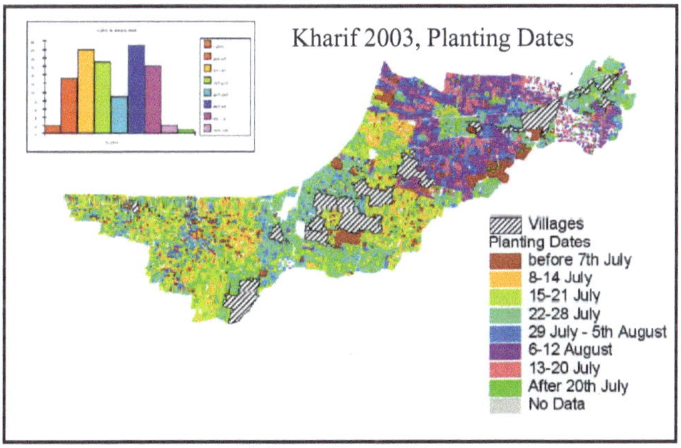

Map Page 311

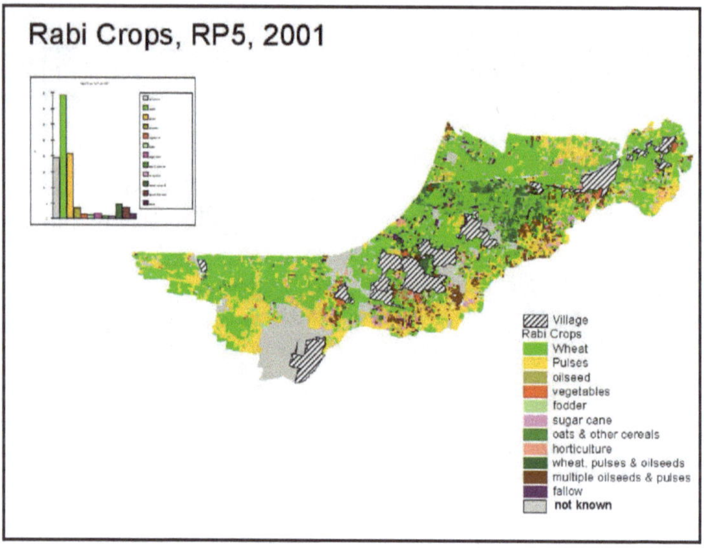

Map Page 312

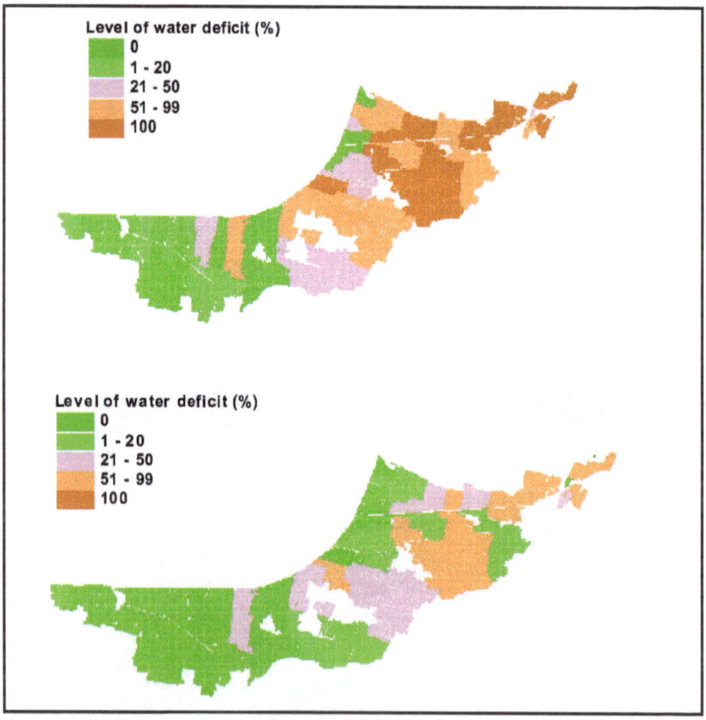

Map Page 320

FSC
www.fsc.org
MIX
Papier | Fördert
gute Waldnutzung
FSC® C083411

Zeitfracht Medien GmbH
Ferdinand-Jühlke-Straße 7
99095 Erfurt, Deutschland
produktsicherheit@kolibri360.de